A Really PRACTICAL HANDBOOK of Children's Palliative Care

for Doctors and Nurses
Anywhere in the World

JUSTIN AMERY

And INCLUDING
(By kind permission of Dr. Sat Jassal)

The APPM Master Formulary 2015

Copyright © 2016 Justin Amery.

All rights reserved. No part of this book may be reproduced, stored, or transmitted by any means—whether auditory, graphic, mechanical, or electronic—without written permission of both publisher and author, except in the case of brief excerpts used in critical articles and reviews. Unauthorized reproduction of any part of this work is illegal and is punishable by law.

The information, ideas, and suggestions in this book are not intended as a substitute for professional medical advice. Before following any suggestions contained in this book, you should consult your personal physician. Neither the author nor the publisher shall be liable or responsible for any loss or damage allegedly arising as a consequence of your use or application of any information or suggestions in this book.

ISBN: 978-1-4834-4402-4 (sc)
ISBN: 978-1-4834-4403-1 (e)

Library of Congress Control Number: 2015921132

Because of the dynamic nature of the Internet, any web addresses or links contained in this book may have changed since publication and may no longer be valid. The views expressed in this work are solely those of the author and do not necessarily reflect the views of the publisher, and the publisher hereby disclaims any responsibility for them.

Any people depicted in stock imagery provided by Thinkstock are models, and such images are being used for illustrative purposes only.
Certain stock imagery © Thinkstock.

Lulu Publishing Services rev. date: 02/19/2016

Contents

A Word of Thanks ix
Foreword xi
Contributors xiii
How to use this book xxi
Part one: How do I communicate with children and their families? 1
Part two: How do I break bad news to children and their families? 9
Part three: How do I handle strong emotions and difficult questions? 17
Part four: How do I assess and manage all the needs of a
child and their family? 22
Part five: How do I manage children with HIV&AIDS in CPC? 30
Part six: How do I manage symptoms in children's palliative care? 44
    How do I assess pain and distress in children? 44
    How do I manage pain in children? 49
    How do I manage fluids and electrolytes in children? 61
    How do I manage feeding problems in children? 64
    How do I manage sore mouth in children? 67
    How do I manage dysphagia in children? 69
    How do I manage gastro-oesophageal reflux 72
    How do I manage gastro-intestinal bleeding in children? ... 74
    How do I manage nausea and vomiting in children? 77
    How do I manage constipation in children? 83
    How do I manage diarrhoea in children? 87
    How do I manage cough in children? 90
    How do I manage haemoptysis in children? 93
    How do I manage breathlessness in children? 96
    How do I manage seizures in children? 100
    How do I manage spasticity and dystonia in children? 103
    How do I manage skin problems in CPC? 108
    How do I manage anxiety in children? 119

Part seven: How do I manage acute, distressing terminal symptoms at the end of life? 123
    A practical approach to treating any acute, distressing terminal event in CPC: 128
    How do I manage restlessness and agitation at the end of life? 129
    How do I manage massive bleeding at the end of life? 129
    How do I manage acute severe upper airway obstruction (choking) at the end of life? 130
    How do I manage noisy secretions ("death rattle") at the end of life? 131

Part eight: How do I play with children? 133

Part nine: How do I provide palliative care for adolescents? 141

Part ten: How do I deal with ethical dilemmas in children's palliative care? 150

Part eleven: How do I offer spiritual care to families? 157

Part twelve: How do I provide good end of life care to a child and their family? 163

Part thirteen: How do I deal with the practicalities arising after the death of a child? 174

Part fourteen: How do I help the family with grief and bereavement? 179

Part fifteen: How do I survive and thrive in children's palliative care? 187

APPM Formulary 197
Appendices 343
    Appendix 1: Morphine equivalence single dose [1, 2, 5] 343
    Appendix 2: Subcutaneous infusion drug compatibility 344
    Appendix 3: Template Symptom Management Plan 345
    Pain Management Plan Template 346
    Nausea and Vomiting Management Plan Template 348
    Seizure Management Plan Template 350
    Acute Breathlessness Management Plan Template 353
    Excessive Respiratory Secretions Management Plan Template 356

  Severe Bleeding Management Plan Template 359
  Muscle Spasm (Dystonia) Management Plan Template 362
  Mouth Pain Management Plan Template 365
  Spinal Cord Compression Management Plan Template 368
  Agitation, Restlessness, Delirium and
  Anxiety Management Plan Template 371
  Constipation Management Plan Template 373

Formulary References 375
Practical Handbook References 401
Main Index 409
Formulary Index 415
Drug Index 421

A Word of Thanks

I would like to thank the following organisations without whom we would not have been able to publish this book for free.

- Keech Hospice (www.keech.org.uk) who have provided me with logistical, psychological and material support, as well as giving me the freedom to work on the book as part of my commitment to them.
- The True Colours Trust, a charity manned by a small group of people but which has had a huge impact in developing children's palliative care around the world, for their donation to cover the publishing costs of the book.
- The Association for Paediatric Palliative Medicine formulary group, for allowing me to include the formulary in this book for free.
- The International Children's Palliative Care Network (www.icpcn.org) which has been continuously supportive of this work, and which has agreed to help with the dissemination of the book once published.
- Finally to Together for Short Lives (www.togetherforshortlives.org.uk), partly for their contributions to this book, through Katrina McNamara) and also for agreeing to allow me to build on the 'CPC Handbook for GPs'.

Foreword

In my travels practicing and teaching children's palliative care around the world, I have been struck by how children's palliative care is essentially the same, wherever you go. Of course there are many, many differences too, but that's not necessarily a contradiction. Wherever they are, children are still children, and palliative care is still palliative care.

In response to this rather paradoxical state of affairs, I have long dreamed of being able to write something that is freely accessible to everyone and which could be just as useful to a doctor in the USA as to a nurse in Bolivia and a to Clinical Officer in Uganda. This book is an attempt to create that, and it builds on what I have seen: that most doctors and nurses already have most of the knowledge, skills and attitudes they need to be good at children's palliative care, but they just lack the confidence or feel frightened by some of the challenges that inevitably occur when a child is dying.

So the book is structured around the common questions that we ask in tricky palliative situations, building on the research, teaching and training I have done over the last 20 years. Each chapter deals with one question, and starts with reminding us what we already know, before going on to fill in some common gaps. Finally, the book ends with a copy of the Association of Paediatric Palliative Medicine (APPM) Formulary, which is a truly excellent piece of work donated very kindly by Sat Jassal and all at the APPM. I hope this will give you everything you need to manage real life cases in real life settings, with confidence, wherever you are.

Contributors

In order to ensure this book is relevant to any doctor or nurse anywhere in the world, I have relied very heavily on contributors who are highly experienced in children's palliative care from five continents (I couldn't find anyone from Antarctica who fitted the bill). They have been invaluable and I offer them my huge thanks. They are as follows:

Dr. Anthony Herbert
Anthony is a paediatrician specialising in paediatric palliative care at the Lady Cilento Children's Hospital, Brisbane, Queensland, Australia. He spent part of his fellowship working at the Department of Pain Medicine and Palliative Care at The Children's Hospital at Westmead, Sydney. He has been working within the Paediatric Palliative Care Service in Brisbane since 2008 and is a Senior Lecturer at the University of Queensland. Clinical and research interests include pain management, insomnia and the use of telehealth

Francis Edwards
Francis has been working clinically for the last three years at Bristol Royal Children's hospital as the Paediatric Palliative Care Liaison Nurse developing services for babies, children and young people's palliative care services. Prior to moving to Bristol he worked in the North of Scotland as a Nurse Consultant in Paediatric Palliative Care. Francis has a wide experience of working in CPC in acute, community and hospice settings as well as being an Honorary Lecture at Aberdeen & Coventry Universities. He has spent a great deal of his career working as a Clinical Nurse Specialist in Paediatric Oncology, at the children's hospital in Brighton and the Royal Marsden in London. Working in paediatric oncology forms his approach to paediatric palliative care. He appreciates and is committed to the importance of holistic and family-centred care, and sees 'soul work' as a key element of care. Francis has also been involved in developing the Care Pathways with the team at TfSL's.

Dr. Jane Nakawesi
Jane is a paediatrician who has been working at Mildmay Uganda for the last 6 years. She completed her Master of Medicine in Paediatrics and Child Health from the Makerere University in Uganda in the year 2007. She received training in Paediatric Palliative Care (PPC) from South Africa in 2009. Since then she has been training both national and international health workers on both short and long courses in PPC. One of her mentors is a renowned CPC specialist, Dr Justin Amery. The initial training at Mildmay with support from the Diana Princess of Wales Memorial Fund (DPWMF) was a 6 months course. Through her leadership and with support from the DPWMF this course was upgraded to a diploma course. The course commenced in April 2014 at Mildmay Uganda. Jane is also keen in conducting operational research and has presented several papers on PPC at both local and international conferences. This year one of her papers i.e. Palliative Care Needs of HIV Exposed and Infected Children admitted to the Inpatient Paediatric Unit in Uganda has been accepted for publication by ecancermedicalscience

Dr Julia Ambler
Julia graduated from UCT with her MBChB in 1998. After her junior doctor years she spent 6 years in Oxford, UK returning home in early 2008. Whilst in the UK she trained and worked as a general practitioner and a children's hospice doctor at Helen and Douglas Houses in Oxford. She is a member of the Royal College of GP's in the UK and holds the Diploma in Paediatric Palliative Medicine from Cardiff University, Wales. She also holds a Diploma in Child Health. From 2008 to 2012, Julia worked for the Bigshoes Foundation in Durban. When This NPO closed in December 2012, she was instrumental in launching Umduduzi – Hospice Care for Children to continue the work in children's palliative care. Through this NPO she consults to state hospitals and hospice programs and trains in children's palliative care. She is also a part-time lecturer in the Department of Paediatrics, Nelson Mandela Medical School, UKZN .

Katie Hill
Katie graduated with a First Class Honours degree in Children's and General Nursing from Trinity College Dublin, Ireland in April 2012 and

received a Gold Medal from Trinity College for her academic excellence. Throughout her training she developed a passion for paediatric palliative care and in October 2011 she won the 'Undergraduate Award of Ireland and Northern Ireland for Nursing' for her paper on paediatric palliative care, which has since been modified and published in the British Journal of Nursing entitled "Paediatric Palliative Care in Ireland and the UK". Upon graduating Katie spent six months volunteering in the 'Butterfly Children's Hospices', China and has been involved ever since. She has spent over 18 months volunteering with Butterfly Children's Hospices in China where she currently has the role of head nurse and leads the nursing team across both hospices. Currently she splits her time between China and Ireland, where she is undertaking a Masters in Nursing in Child Health and Wellbeing. It is an absolute privilege to be working with Butterfly Children's Hospices, providing loving palliative care and treatment to children with life-threatening illnesses in China.

Katrina McNamara
Katrina McNamara is Director of Practice and Service Development, with Together for Short Lives, the UK Children's Palliative Care organisation, leading a team aiming to support professionals and organisations in developing best practice and services provision and providing information and support services to children and young people with life-shortening conditions and their families. In 2009, she was awarded the Elisabeth Kübler-Ross Award for Outstanding Contribution for her vision in developing the UK's first Integrated Care Pathway for children's palliative care. She is a member of the International Children's Palliative Care Network Advocacy Committee, focusing on the development of children's palliative care services across the world. She worked as a children's nurse, general nurse and health visitor in Merseyside and the Department of Health in England before moving to work in the voluntary sector.

Dr. Khaliah Aesha Johnson
Khaliah is a pediatric palliative care clinician practicing at Children's Healthcare of Atlanta, Atlanta, GA, USA. Dr. Johnson completed her medical training at Stanford University School of Medicine in Palo Alto, CA prior to completing general pediatrics training at Johns Hopkins

University in Baltimore, MD. Her passion and commitment to caring for children with complex, life-limiting medial conditions lead her to pursue pediatric palliative care fellowship training at the Children's Hospital of Philadelphia, which she completed in 2012. Dr. Johnson has had the opportunity to explore and address the palliative care needs of children and families in several international settings, including Ethiopia and South Africa. Her specific career interests include improving access to community-based pediatric palliative care in resource-limited settings and pediatric hospice care.

Dr. Kuan Geok Lan
Dr. Kuan Geok Lan has been advocating for CPC in Malaysia since 2006, when she set up the first CPC team in Malacca Hospital. She continues to serve in the Dept. of Paediatrics, Malacca Hospital as a general paediatrician and looking after children with CPC needs in the community, as well as teaching medical students. She chairs the Malaysian Paediatric Association, CPC subcommittee, which has led to the recognition of development and training of paediatricians in CPC by the Ministry of Health, Malaysia. The subcommittee also aims to continue education amongst healthcare workers, enhance public awareness, encourage research as well as engage the public. Since 2013, she has been instrumental in setting up a CPC service in two government hospitals in the state of Johore, Malaysia. In 2015 she engaged with 12 rural health clinics supported by these two hospitals in collaborative work, to provide home care and support using a " pop up and on time training" model.

Maggie Comac
Maggie Comac is a Clinical Nurse Specialist in Paediatric Palliative Care at Great Ormond Street Hospital. She has a wide experience of paediatric oncology and palliative care both in the UK and in the Middle East. In June 2010 – May 2014 Maggie worked in Kuwait both within the Ministry of Health as a Nurse Consultant and latterly as Director of Care at Bayt Abdullah Children's Hospice, the first paediatric palliative care facility in the region. Maggie has passion for advanced nursing practice, supporting quality evidenced based care. Her special interest lies within

clinical research within the field and hopes to contribute to a much needed evidence base in her future work.

Dr. Pradnya Talawadekar
Pradnya is the Country Co-ordinator of the Children's Palliative Care Project, Indian Association of Palliative Care. She has helped advocate for the implementation of palliative care in the State of Maharashtra and the inclusion of palliative medicine into the curriculum of Maharashtra University of Health Sciences (MUHS). She has developed educational & advocacy material in India and organised training for Indian health care worker. She works with NGOs to provide help and advocacy for children with palliative care needs and has helped develop three model sites for service delivery in Maharashtra'

Dr. Renee McCulloch
Renée is a Consultant in Paediatric Palliative Medicine at Great Ormond Street Hospital; Honorary Senior Clinical lecturer at the Institute of Child Health, London and also works at Helen and Douglas House Hospice in Oxford. She has a broad background that includes working in specialist paediatric centres, children's hospices and delivering palliative care at home. Between 2010 and 2013 she worked in Kuwait supporting the development of children's palliative care services at the children's cancer centre and then at Bayt Abdullah children's hospice. Between 2010 and 2013 Renée and her family went to live and work in Kuwait supporting the development of children's palliative care services. Renée worked initially in the children's cancer centre and then at Bayt Abdullah children's hospice (the first children's hospice in the Middle East). Renee has contributed widely to the specialty of children's palliative care and enjoys being involved in research, training and education. She is particularly interested in developing an evidence base for practice, improving children's palliative care services and increasing the availability of opioid preparations for children so they are accessible in all countries.

Dr Rut Kiman
Rut is the head of the Pediatric Palliative Care Team in the Maternal-Infant Department at the 'Professor A. Posadas Hospital' in Argentina

as well as Assistant Professor of Pediatrics and Senior Lecturer in the Pediatric Department of the School of Medicine, University of Buenos Aires. Rut has been working with children and families for more than 30 years, and 20 years of experience dealing with infants and children with life-limiting conditions. She also teaches undergraduate students, medical residents in Posadas hospital and postgraduate students in the Diploma Course of NGO, Pallium. As a member of the Latin-American Palliative Care Association, she has participated in many workshops in different parts of that region. Board member of International Children Palliative Care Network since 2006, she has also contributed with several articles and book chapters in her speciality.

Dr. Satbir Jassal
Sat is a full time senior partner general practitioner and Medical Director of Rainbows Children's Hospice in Loughborough, Leicestershire heading a team of 4 doctors on a 14 bedded unit that deals with children and young adults. He has been a GP for 25 years and have worked in children's palliative care for over 20 years. He has written and edited the Rainbows Children's Hospice Symptom Control, co-authored the Oxford Handbook of Paediatric Palliative Medicine and is an editor of Synopsis an abstract journal. He is chair of the APPM Drug Formulary Group and editor of the APPM Master Formulary; as well as lecturer at Cardiff University in Paediatric Palliative Medicine and Honorary Consultant for LOROS (an adult hospice). He is the Child Health Lead for West Leicestershire CCG and Clinical lead for EOL for the East Midlands Strategic Clinical Network. He has been awarded an honorary FRCPCH and in 2014 a MBE.

Sue Boucher
Sue Boucher is an educator, consultant editor and author of children's books and educational textbooks. She edits the International Children's Edition of e-Hospice and is responsible for social media for the ICPCN. She is an Advisor to the Elizabeth Kubler-Ross Foundation, sits on the steering committee of Palliative Treatment for Children in South Africa (PATCH – SA) and is a board member of Umduduzi Hospice Care for children in Durban. Sue joined the ICPCN in 2007 as the International Information Officer and is now Director of Communications. She is

active in education, advocacy, development of materials for the ICPCN, and contributes to textbooks, journal articles and conferences.

Dr. Zipporah Ali
Zippy is the Executive Director or of Kenya Hospices and Palliative Care Association (KEHPCA) and board member of several other international CPC organizations. She was among several others who received the African Palliative Care Association (APCA) award for her contribution to palliative care in Africa and her work on the inaugural board of APCA. She has previously served on the APCA board and The Kenya Cancer Association Board. She is involved in advocacy and creating awareness on palliative care in Kenya and has been instrumental in integrating palliative care into Kenyan government hospitals and undergraduate medical and nursing schools curricula in Kenya. Zipporah was awarded an Honorary Doctorate of the University by Oxford Brookes University in recognition of her outstanding contribution in palliative care both internationally and locally; an advocacy award by the from the African Palliative Care Association and Open Society Foundations.

Other Contributors
We are also very grateful for the contributions of Lyn Gould (Butterfly Children's Hospice in China) and Dr. Martha Mherekumombe (Department of Pain Medicine and Palliative Care at Sydney Children's Hospitals Network)

How to use this book

Hopefully you will find it easy to use this book. Each chapter is based around a question stem:

'How do I….?'

So simply go to the contents page, choose the problem that most closely fits the problem you are dealing with and follow the advice.

If the advice involves using drugs, go to the APPM Formulary at the back of the book and look the drug up. The APPM formulary is the best CPC formulary there is, and it is constantly being revised and updated.

If you would like to read more around the subject, then there are suggestions for you at the end of each section.

References can be found at the very back of the book.

Out of habit I have used UK English rather than American English spellings. Please forgive me if you prefer the latter. Similarly, I have used generic names of drugs, using UK spellings if there is any difference internationally.

Part one

How do I communicate with children and their families?

What you probably know already

- You probably know more than you think about communicating with children. Even if you don't have children in your family, you can draw upon your own experience of childhood.
- Doctors and nurses are trained in communication, especially in the use of non-verbal communication and communicating with families. These skills transfer easily into the children's palliative care field.
- There are lots of things beyond your control. As a doctor or nurse, a big part of your work is to help people live with problems they cannot change.
- Dealing with the death of a child can be quite frightening and emotionally draining without proper systems and structures to support you, but done properly, it can be a highly rewarding aspect of your job.
- Good children's palliative care requires good team working, which is often a great deal harder than it sounds

What you might find useful

Listening
In all areas of doctors' or nurses' work, good communication is essentially about deep listening, and is sometimes hearing what might not be said. Children may never say what is wrong or what is on their mind.

How do children communicate?

- **Body language**: Children are often easy to read, but can also really fool you sometimes. They are excellent readers of your body language, so be aware of what your own body language might be conveying to the child and adopt an open, friendly and relaxed posture.
- **Play language:** Children will often show you rather than tell you about their life. Keep a few play materials handy, set aside some time, sit down next to a child, look interested and wait. They will do the rest.
- **Spoken language:** Speaking is not the first language of very small children, and even older children may prefer to converse in other ways. Imagine you are speaking to someone from a different country, with a different culture and a different language, and be sure to filter out jargon and presuppositions. For older children who can speak in their first language, you may need a translator to interpret if your first language is not the same as the child.

Why is communication so important in children's palliative care?

- By listening and responding, we may discover what children know and do not know. We can then help by providing information, comfort and understanding.
- Not talking about something doesn't mean we are not communicating: avoidance in itself is a message.
- Children and parents tend to protect each other from being upset by avoiding difficult discussions, so the child can at times become emotionally isolated.
- Good communication helps children to become involved in their own care management[1] and can improve adherence to treatment.
- Good communication will also reduce complaints or dissatisfaction about the care provided to patients

- Open communication with children and their families may improve your professional job satisfaction.
- And (of course) communication helps us to elicit useful information so that we can diagnose problems and develop good care management plans.

What are the barriers to good communication with children?

Societal factors: In resource-rich society, childhood death is now rare, whereas in resource poor settings it is still terribly common. People in settings where childhood death is rare may lack experience, be fearful of handling death or may have unrealistic expectations. On the other hand, people in settings where childhood death is common may become 'immune' to it and so really struggle to deal with it effectively and sensitively.

Cultural factors: Cultural factors play a major role in how we view health, illness and dying and so it is vitally important that we are both aware of our how own cultural 'goggles' may distort our world-view, and also that we respect and consider the cultural world-views of others. Please note, there is a very big 'BEWARE!' here. We often tend to assume that children share the same culture as their families. However, while true in part, 'child-world' is a very different place to 'adult-world', whatever the parents' culture may be. Cultural awareness and respect is more than being polite to people who may be different to ourselves (important though that always is). It has practical implications too, because cultural factors affect the way we think and behave as parents and as professionals, and therefore have a profound impact on how a child experiences death and dying..

Language factors: Language barriers can create difficulties when trying to communicate effectively. It is important the family, child and health care professional all understand what is being said. A translator may be necessary for accurate interpretation.

Patient factors: Children and their families may under-report problems because they fear nothing can be done, because they fear burdening carers, or because they wish to appear strong. Children often under-report their problems, partly because they fear going to hospital and/or painful procedures, partly because they have got used to living with the problem, and partly because they know that difficult discussions upset their parents, which they do not want.

Making assumptions: We often make unsafe assumptions about children, for example, assuming a child with advanced cancer is worried about the pain during the end of life stages, when in fact she may (for example) be terrified of worms eating her body after death. Honest, open communication should help to avoid making assumptions.

Distancing: Caring for individuals who are a constant reminder of your own worst fears (and maybe memories) can be very difficult. To cope, sometimes professionals may avoid caring for such patients altogether or avoid having difficult conversations which need to happen.

Fear of doing harm by causing upset: Children with life-limiting conditions usually know much more than parents and professionals think, and there is very little evidence that giving children too much information does any long term harm. Trying to find the right balance between respecting the parents views and trying to work in the best interests of the child is a tricky, but essential part of CPC. Collusion and mutual pretence may often lead to decisions about a child being made without engaging the child or finding out his/her concerns, ideas and expectations. Being 'locked in' like this is very damaging for a child and makes good CPC very difficult indeed.

Fear of provoking strong emotions: Anger, crying, denial and depression are some of the emotions and reactions that we associate with the death of a child. Most of us do not enjoy being on the receiving end of these emotions, and it is natural that we try to avoid them. It is important to allow family members and the child, space, time and opportunity to express these feelings.

Perception of lack of skills: You may feel you could use more skills, but be encouraged that it is very unlikely that you will do any harm by talking to children about difficult issues. It is very likely that you will do a lot of good by listening and just being able to remain with a child during a difficult and painful time, even if you can't take the pain away.

Sympathetic pain: We all fear illness and mortality, and have all suffered painful losses. We are taught to maintain a professional façade, so fear of losing control and expressing our own emotions can be an even bigger block to communication.

Fear of having no solutions: You may feel that you ought to have a cure or a solution for a patient's problems, or feel you have failed if you don't. However the real myth is thinking that there are solutions to life and death in the first place. While you won't be able to prevent death, there are many ways in which you can help.

How do I go about communicating with children?

Just because you haven't said the words, it doesn't mean you haven't expressed something. Remember the power of non-verbal communication. If you or the family are avoiding a thorny issue, remember that the child is likely to have picked up that there is something that is too bad or frightening to talk about. Try to imagine how that feels.

Take time: Remember the old adage: 'more haste, less speed'. Ask yourself: Do I need a trained translator as opposed to a family member of friend? Is this the right environment? Will I be disturbed?
Tell your colleagues or practice that you are out of contact, turn off the phone handover your mobile to a colleague, sit down, try to consciously relax your body and open your expression. The child needs to know that at that given time they are your only priority.

Connect: Use an open and friendly manner, take time, shake hands, introduce everyone, explain your role, and generally establish ease and rapport with both the child and the family.

Look at yourself: You may have forgotten what the world looks like when you are two feet tall. Try sitting on the floor while the adult world buzzes around you. To a child you probably look seriously scary. So smile, make yourself smaller, get down to their level, crouch up, and avoid big, fast, scary movements. Being aware of your own feelings (including your breathing) can sometimes help you to focus more on the child and parent (a form of mindfulness).

Respect space: Respect the child's personal space. Wait until they come to you, or at least wait to be invited in. Don't be a cheek-pincher, an arm-puncher, a head-patter or a hair-ruffler. **Touch:** Once you are allowed closer, touch can be acceptable, or it can be intolerable to a child. The problem is, sooner or later you are going to have to examine the child, so you need to break the ice at some time. A very gentle touch on the hand or arm can speak volumes to a child.

Make the environment child friendly: Do you have pictures of nasty diseases on your walls? Take them down and replace them with something more soothing. Bright, coloured rooms make the child feel more comfortable and relaxed. Probably the best is to get the children you see to draw you some pictures or make you some models that you can stick on the wall. Are there frightening instruments or equipment on display? Do you have any toys or crayons in your room?

Ensure the child is comfortable: Make sure the child is comfortable (e.g. on Mum's knee, on the floor playing with toys, on a bed etc.). Does she need to be examined on the couch, or will Mum's lap do? What is wrong with examining on the floor if the child is playing there happily anyway?

Offer regular support and praise: At every opportunity make sure you say *"well done"* or *"good girl"*, even if it is just for coming into your room without crying. The more you build a child's confidence, the more co-operative he or she will be. It can be useful to have things like stickers or lollipops close by, to reward a child's good behaviour, again gaining trust while doing so.

Explain what you are doing: If you need to undertake procedures that will be painful to the child explain what you are doing and that it may hurt. Where possible, use distraction techniques or drugs to prevent and/or manage it. Whatever you do, don't be tempted to tell them that it will not hurt because if it does they will lose trust in you and will be fearful of all health care professionals who they subsequently come in contact with.

Show respect: Avoid being patronising, judgmental, harsh or brusque, particularly with older children. Listen and attend to everything that they have to say, don't make promises you can't keep, and be true to your word. Ask the adolescent if they wish to have someone with them whilst you talk or undertake a procedure. Do not interrupt them, allow them to speak until they have finished what they wanted to say. For adolescent females always have a female chaperone with you.

Avoid assumptions: Always ask questions in response to the child's questions in order to clarify understanding and the meaning behind some questions. For example, to the question *"Am I dying?"* you may respond *"what makes you think that?",* or something similar. No two children are the same, so do not assume because you have dealt with a similar situation before that this child would also feel the same.

Have a go: If you mess it up, don't be put off. If you have the right intentions, you will be able to regain rapport. Saying sorry and asking for another chance tells children you respect them, and may even put you in a better position than you started from. Children are usually incredibly forgiving and are prepared to give you a second chance (and a third, fourth, fifth….) as long as they trust you and know that your efforts are for the best.

What other resources might I find helpful?

- Darnill, S. and Gamage, B. (2006) 'The patient's journey: palliative care – a parent's view'. British Medical Journal [online] 332 : 1494 doi: 10.1136/bmj.332.7556.1494 <http://www.bmj.com/content/332/7556/1494.full>

- Ranmal, R., Prictor, M, Scott, J.T. (2008) 'Interventions for improving communication with children and adolescents about their cancer'. Cochrane Database of Systematic Reviews 2008, Issue 4. Art. No.: CD002969. DOI: 10.1002/14651858.CD002969.pub2. Available from <http://onlinelibrary.wiley.com/o/cochrane/clsysrev/articles/CD002969/frame.html>
- Amery, J. (Ed.) (2009) Children's Palliative Care in Africa. Oxford: Oxford University Press. Available from <http://www.icpcn.org.uk/core/core_picker/download.asp?id=204>
- Goldman, A., Hain, R. and Liben, S. (2012). (Oxford Textbook of Palliative Care for Children. Second Edition . Oxford. OUP.
- gp-training.net (2006) Communicating with Children [online] available from <http://www.gp-training.net/training/communication_skills/consultation/children.htm>
- ICAN http://www.ican.org.uk/
- Kids Behaviour (UK): http://www.kidsbehaviour.co.uk/ A website offering advice, help and support to parents or carers who need guidance when dealing with a child's behaviour
- Levetown, M. (2008) 'Communicating With Children and Families: From Everyday Interactions to Skill in Conveying Distressing Information'. *Pediatrics,* 121(5), pp.e1441-e1460. [Online]. Available from: http://www.unboundmedicine.com/medline/citation/18450887/Communicating_with_children_and_families:_from_everyday_interactions_to_skill_in_conveying_distressing_information_
- Osbourne, H (2001) ' In other words…Start Where They Are…Communicating With Children and Their Families About Health and Illness' *On Call Magazine.* [Online]. Available from: http://www.healthliteracy.com/article.asp?PageID=3776

Part two

How do I break bad news to children and their families?

What you probably know already

- The principles of breaking bad news are the same in any situation, whether with children or adults.
- Psychological defence mechanisms have a profound impact on our ability to handle and communicate distressing information.
- Doctors and nurses have to handle lots of potentially distressing information, and are usually highly competent at judging how best to do this.
- People usually come to their own realisation that they are going to die, or that their child is going to die, and often require little more than gentle and consistent support for this realisation to occur.

What you might find useful

- Palliative care is much easier, more effective and more supportive when everyone is aware and everyone is communicating.
- In practice, this perfect state of affairs is not always easy to achieve.
- When a child dies, the child and family have to make two journeys: the physical journey to death, and the emotional journey to death.
- As long as the emotional journey does not lag too far behind the physical one, there is usually no need to push children and families towards awareness of death. You have time to try and build trust and communication.

- However, sometimes the emotional journey lags so far behind that it becomes impossible to make effective plans or decisions. When this happens, children can suffer unnecessarily.
- Sometimes it is our colleagues that block disclosure, not from bad intent, but because they themselves are not ready. When this occurs we can find ourselves stuck in a tricky triangle.
- At some point, preferably sooner rather than later, you are going to have to disclose the child's prognosis to the family, and ideally to the child.
- Doctors' and nurses' - perfectly understandable - fear of this disclosure is arguably the biggest reason (at least in my experience) why CPC is often done so poorly around the world.
- So you REALLY need to be able to do it…

Key goals in breaking bad news:

- Parents should be treated with openness and honesty.
- Parents should be acknowledged as experts in the care of their child.
- Significant news should be shared in a place of privacy.
- Professionals should allow plenty of time for sharing news and discussing what this means with families.
- Parents should be given the opportunity to hear news together. Advocates and interpreters should be readily available to support families.
- News should be shared using clear, jargon-free and readily understandable language.
- There should be open communication between professionals and the family.
- Parents should be given time to explore care options and ask questions.
- Breaking significant news should be backed up by helpful written material as the child and family often forget a lot of what has been said, due to shock and stress..

8 steps to breaking bad news

Actually breaking the news is difficult. Here are some key steps in the process to help you:
1. Prepare for breaking significant news.
2. Assess the awareness of everyone involved.
3. Find out how much the child and family know.
4. Find out how much the child and family want to know.
5. Allow time for silence.
6. Manage denial and collusion.
7. Break the news using the Warn Pause Check (WPC) approach.
8. Respond to the child's and family's feelings.
9. Plan and follow through.[2]

1. Prepare for breaking significant news
- You should prepare yourself in advance. Make sure you have obtained and grasped everything you can about the patient's condition and management to date.
- Anticipate the kind of questions you might be asked, and think about what you would ask in this situation.
- Practice speaking phrases and sentences in advance. Don't just think about them, but actually practice them, preferably in front of a mirror. This will give you the confidence you need in order to speak to the family.
- Plan the location where you will break the news. Make sure it is a relaxed environment, giving the family complete privacy, and ensure the right people are present.
- If you suspect that different family members have different levels of knowledge or are approaching the situation very differently, it might be appropriate to see them separately. However, make sure you get to all the key decision makers in the same time frame, or you will risk causing tensions and conflict.
- It may or may not be appropriate for the child to be there. Often, and particularly with smaller children, parents tend to prefer

having the news broken to them first, and then taking part in breaking significant news with the child themselves.
- The degree to which parents should have a 'veto' over what information is to be given to the child seems very variable across the world, with different contributors sharing different opinions, although all agree the ideal situation is for open disclosure and discussion wherever possible.

2. Assess the awareness of everyone involved

When a child and family are confronting a life-limiting or life-threatening diagnosis, it is possible for them to be in one of four different 'awareness contexts':

- **Closed awareness**: The child is not aware of the diagnosis, and those who know conceal it.
- **Suspected awareness**: The child is suspicious something is wrong, but is not certain.
- **Mutual pretence**: The child, family and health workers all know, but no one talks about it.
- **Open awareness**: Everyone knows and is open about it.

Open awareness is the ideal as it allows for fears and concerns to be voiced and addressed, for better care plans to be negotiated and agreed; and for the child and family to feel more in control.

Often child, family and professionals are each in different awareness contexts, or stuck in 'mutual pretence'[3], usually because the truth is too distressing or difficult to handle.

Generally it is fine to allow everyone to reach open awareness in their own time, but where blocked communication risks increasing a child's suffering (e.g. where a child is isolated and anxious, or where events are proceeding so fast that communication is crucial to plan, prepare and adapt), then you may need to intervene (see 'Managing Denial and Collusion' below).

3. Find out how much the child and family know

Try and find out how much each person knows. Ask each one individually and try to prevent others blocking or interrupting Always allow people to speak freely to allow them express what they really want to say. Take time to listen whilst also allowing silence and space.. When they speak, reflect back what they have said and make sure you have understood exactly what they know before moving on.

4. Find out how much the child and family want to know

First you need to find out the level of denial the child or family members have, either consciously or subconsciously. If they signal they are not ready to be open, back off and review this later on. Don't push information if they don't want it.

5. Manage denial and collusion

This is an art, not a science, so there are no clear, 'one size fits all' answers. You need to try and balance the risks and the benefits of allowing the denial to continue. Where the motivation is love (rather than control), collusion and denial tend to melt away as events progress and as people adapt to the situation. In this case you might just need to wait until this happens naturally. Where things are deteriorating fast and time is not a luxury you have; or where the child is clearly being isolated and upset by the collusion or denial, you need to explain to them all that you recognise that everyone is acting as they are, purely to prevent others being hurt. But also explain that by not allowing communication, they may be inadvertently hurting their child. Children can pick up on non-verbal behaviours, other peoples' emotions and will very quickly be able to tell if there is tension and stress amongst adults.

6. Break the news using the WPC Chunk Method

A good way is the 'WPC Chunk' method. It is simple and it works.
Start off by mentally breaking the news into chunks. With a life-limiting condition, there are several 'chunks' of news that need to be imparted to the child and family:
- That he or she will die.
- That he or she will die soon, or may have a long-term condition with slow degeneration and complex care needs.

- That he or she might suffer with unpleasant symptoms unless carefully managed.
- That the family will need to learn what to do if any of these eventualities arise.

Few people, if any, would be able to absorb all of these chunks of information in one go. People automatically cut out after a certain level of pain and stop hearing. You might want to get it all over and done with, to get the significant news out all at once. That's understandable, but chances are, it won't work.

The 'WPC Chunk Method'
Decide which 'chunk' of news you are going to try to share.
- **Warn**: This gives the person a chance to prepare and brace themselves, and helps them to absorb the difficult information.
- **Pause**: This gives the person a chance to decide whether or not they still want to go ahead, and also to react. If they assent (either verbally or non-verbally), go ahead and share the first chunk of news.
- **Check back**: Ask what they have understood and correct or reinforce. This allows you to ensure they have understood you, and also acts to embed the news properly in the person's memory. This is important because, after traumatic events, people often forget what happened and what was said.
- Decide whether they are ready to move on, and if they are, then share the next chunk of news using the same method.
- At all stages, written information should be provided for reinforcement for future use.

7. Respond to the child's and family's feelings
Now just sit and wait for the child and family to react. Whatever their reaction, stay calm. If you start to become emotional or fearful, try not to panic. It will fade and pass as long as you don't inflame it. Be gentle and show through your non-verbal communication that you have all the time in the world (even if you don't). Most importantly, whatever the response

is, validate it. You might say something like, *"it's OK to be angry/upset"* or *"many people find it difficult to speak after hearing news like that"*.

Once the response settles, you should repeat the process until one of four things happens:
1. There is no more news to share.
2. They signal that they have had enough.
3. You get the feeling that they have stopped hearing or absorbing.
4. You feel you personally cannot do anymore (which is fine, as long as you make sure you arrange to come back).

8. Plan and follow through
Once all the news is out; you have allowed time for individuals to react and express their emotions; and you have validated them, move from listening mode into a slightly more active mode. Begin to identify options, suggest sources of support and start negotiating care management plans for the various problems and issues that you have identified. Provide written information and leaflets for them to take away and read over the next few days. Whatever else you do, make sure the family is able to make contact with you or a colleague over the next few days. Ensure they have contact details on hand, to make them feel more comfortable and supported during this time.

What other resources might I find helpful?

- Buckman, R.A. (2005) *Breaking bad news: the S.P.I.K.E.S. Strategy.* Community Oncology. <http://www.communityoncology.net/journal/articles/0202138.pdf>
- Goldman, A. Hain, R. and Liben, S. (2006) *Oxford Textbook of Palliative Care for Children – Chapters 2, 7-10.* Oxford: Oxford University Press
- Buckman, R.A. (1992) *How to Break Bad News: A Guide for Health Care Professionals.* Baltimore: John Hopkins University Press.
- Department of Health, Social Services & Public Safety (DHSSPS) (2003) *Breaking Bad News ... Regional Guidelines Developed from Partnerships in Caring.* Belfast: DHSSPS

- Scope (1993) *Right From The Start – template document. A guide to good practice in diagnosis and disclosure.* London: Scope Publications
- Proulx, M. (2009) ' Using silence as a communication tool'. *Analytical Mind.* [Online]. Available from: http://analytical-mind.com/2009/11/23/using-silence-as-a-communication-tool/
- Irish Hospice Foundation (nd.) *How Do I Break Bad News – Information for Staff.* [Online]. Available from: http://hospicefoundation.ie/wp-content/uploads/2013/04/How-Do-I-Break-Bad-News.pdf
- Blancaflor, S. (2010) *Communication and End of Life Decision Making – A Resource Guide for Physicians.* [Online]. Available from: http://www.ctendoflifecare.org/documents/CommunicationandEndofLifeDecisionMaking_000.pdf
- RCN (2013) *Breaking bad news: supporting parents when they are told of their child's diagnosis* [Online]. Available from: http://www.rcn.org.uk/__data/assets/pdf_file/0006/545289/004471.pdf

Part three

How do I handle strong emotions and difficult questions?

What you probably know already

- Emotional reactions are part and parcel of living with life-limiting or life-threatening conditions.
- As a doctor or nurse you are well used to facing difficult and often unanswerable questions.
- Your role as a doctor or nurse is to treat what you can treat and help your patients come to terms with what you can't.
- You are used to helping patients cope with their problems.
- You are used to using different non-verbal and verbal tools to assist communication.
- Maintaining your own health and well-being is your obligation to your patients, to your family and to yourself.

What you might find useful

Dealing with strong emotions
When children and families express anger, profound sadness and other strong emotions it can be quite frightening and distressing. However, these reactions are both normal and appropriate when a child is diagnosed with palliative care needs, and may often be helpful and cathartic. The most important thing to remember is that while it may feel as if the emotions are aimed at you, in fact they are directed at the situation. If you find yourself in the midst of powerful emotions, remind yourself that

nobody can maintain them for any length of time. They usually burn out fairly rapidly, even though it may not feel that way. The best things you can do are:

- Ensure you are in a safe environment with access to leave should you need to.
- Open your posture, stay small, and be as empathic as you can.
- Take deep breaths and breathe slowly.
- Listen.
- Don't interrupt or argue.
- Once the emotion has dissipated, cautiously reflect back, check and validate what you have heard to ensure that you understand what is being said.
- Legitimise family members' emotions where possible, and always support.
- Offer to meet again, agree a time and place. If the child or family do not wish to meet again, ensure they have a contact point for you during working hours.
- Make sure to speak to someone about your emotions after any emotional discussion.

Dealing with difficult questions

Children can and do use inopportune moments to ask their questions. They also don't always 'do' adult niceties. Questions from a child can be very blunt and to the point. Chances are, they will ask a tricky question when you least expect it and are least prepared for it.

However, question-asking by a child is a sign of trust, and if a child is prepared to trust you with their concerns and fears, there is a very good chance you can do something extremely helpful and worthwhile, which is to put their mind at rest and help them prepare for their death. Here are some tips that should help get you through most situations:

Answer questions with questions until you are sure you are on the right wavelength: For example, if a dying child says: *"what will happen to me?"* don't launch immediately into death and the afterlife. They may well want to talk about that, but it might be something much more mundane. Answer their question with your own. For example ask: *"That's a good question. But, before I answer, what do you think might happen to you?"* Once, I was answered with *"I think I might miss dinner because I have to wait for my medicine to arrive."*

Don't use your own questions to avoid answering the question the child has asked: They will spot it and you will lose their trust.

Use clear language appropriate to the child's age: Don't use jargon. Remember children may not even know some quite basic anatomical words, like 'chest' for example.

Don't use metaphors or euphemisms: Adults use a whole raft of euphemisms and abstract concepts to talk about death and the afterlife. Abstract or religious language can be extremely confusing and distressing to children. If you describe death as 'sleep', don't be surprised if the child refuses to go to bed or if the child worries that they may not 'wake up' again. A child who is asked if he wants to see his brother's body may wonder why he cannot see the head too. A child who has been told their mother is in heaven might worry why she does not visit or write. It is important to be clear with the child and explain in simple terms with language, which is easy for them to understand and is age-appropriate.

Give simple, clear and honest answers to questions: If you do get into a conversation about death with a child, be completely honest. Children tend not to ask adults until they have exhausted all other possible lines of enquiry (particularly other children, TV, books, the internet, eavesdropping on adult conversation and using their imaginations). The chances are that a child will only ask when one of two situations arises. One, they already know the answer and want you to confirm it. Two, they have a confused or worrying idea of what might happen and they need you to explain. Either way, you need to be honest, clear and straightforward. If you do not, you will lose their trust and/or cause confusion and worry.

Check back after explanations: Once you have finished speaking, always get them to repeat their version of what you have just said. It is amazing how often children get the wrong end of the stick. Or perhaps, more accurately, it's amazing how often we give children the wrong end of the stick.

Be prepared to say you don't know: Children live in a whole world that they don't understand. So children are much more comfortable than adults about not knowing answers to things. Contrary to popular belief, they don't expect adults to know everything (even though they never seem to stop asking questions). If you don't know, say so. Tell them you will try and find out the answer or you will tell them as soon as you know. You will gain trust. If you go on to help them find out the answer, and you will also win a great deal of appreciation. Both will be very helpful as you get closer to the end and you need their co-operation and trust. Recognise that spiritual beliefs affect understanding and hopeful thinking, so verbalise your understanding of individual beliefs, whilst also maintaining medical honesty regarding the clinical situation.

Don't be scared to show emotion, but if you do, make sure you explain it: On the whole children are familiar with and comfortable with emotion, just as long as they understand where it is coming from and that it is not because of anything they have done wrong. If anyone (yourself included) gets upset during the consultation, make sure you take time to explain why they are upset and reassure the child it is nothing they have done wrong or it is not their fault.

What other resources might I find helpful?

- Lo, B., Ruston, D., Kates, L.W. Arnold, R.M., Cohen, C.B. Faber-Langendoen, K., Pantilat, S.Z. Puchalski, C.M., Quill, T.R., Rabow, M.W., Schreiber, S., Sulmasy, D.P. & Tulsky, J.A. (2002) ' Discussing Religious and Spiritual Issues at the End of Life'. *JAMA,* 287(6), pp. 749-754. [Online]. Available from: http://services.medicine.uab.edu/publicdocuments/palliativecare/Discussing%20Religious%20and%20Spiritual%20Issues%20at%20the%20End%20of%20Life.pdf

- Austrailian Palliative Care Organisation (1999) *Multicultural Palliative Care Guidelines*. [Online]. Available from: http://www.palliativecare.org.au/portals/46/resources/multiculturalguidelines.pdf
- Al-Shahri, M.Z. & Al-Khenaizan, A. (2005) 'Palliative Care for Muslims'. *The Journal of Supportive Oncology,* 3(6), pp.432-436. [Online]. Available from: http://www.oncologypractice.com/jso/journal/articles/0306432.pdf

Part four

How do I assess and manage all the needs of a child and their family?

What you probably know already

- Many, although not all, doctors and nurses are well used to looking at patients as individuals within a multi-disciplinary context.
- There are many different factors that come together to influence quality of life.
- Children have ongoing developmental, educational, identity and dependency needs in a way that most adults don't.
- Getting to the root of the problem is only half of good practice. To make a difference, problems need to be managed.
- Care planning is a process of discourse with the child (where possible), family and other relevant people.
- Good care planning involves clarification and agreement on specific and realistic management objectives, contingency planning, and arrangements for appropriate review and 'hand over' to the family.

What you might find useful

If the main reasons for poor palliative care around the world are reluctance of professionals to recognise impending death and to break the bad news, arguably the next biggest reason is poor communication, coordination and teamwork between professionals and agencies.

So, if you are thinking that this chapter looks a bit boring and you might skip it, please don't.

Assessing and planning care

Good children's palliative care planning means hoping for the best and preparing for the worst. Without a full assessment and an agreed, achievable and realistic care plan it is very difficult to provide good children's palliative care. We know that professionals often delay difficult discussions and care planning until it is too late, leaving the child and family to suffer unnecessarily or spend their last days in hospital, rather than at home, where they often want to be.

Collaborative multi-agency working is crucial. Trying to set up collaborative working systems with families and co-workers can be frustrating, and time consuming in the initial stages. Nevertheless, to be effective, it is fundamentally important to work within this collaborative, comprehensive framework.

Overview of needs assessment and planning

Assessment of needs and care planning should start as soon as possible after diagnosis, be ongoing and re-visited frequently. Don't wait until a crisis happens in order to address it: if you prepare for the worst, many crises can be avoided.

Ideally assessment and care planning should be a multi-agency and multi-disciplinary process involving the care team (including yourself), the child, the immediate family, any other family members or statutory services who have a decision-making function regarding the child, and other community or voluntary individuals or agencies involved with the child's care. A meeting should be arranged with the family and as many of the people involved in the child's care as possible, ideally in the family home. The assessment may well not get completed during one meeting. Assessment should be seen as an ongoing process rather than a single event; and depending on the individual family's needs, may take days or even weeks.

The Together for Short Lives Care Pathway[4] describes some key goals regarding needs assessment:
- Children and families should have their needs assessed as soon as possible after diagnosis or recognition.
- A comprehensive and multi-agency approach should be used to avoid the need for multiple assessments.
- Assessment of needs should be in partnership with the family.
- The child or young person should be kept in focus and involved in the process.
- Individuality and ethnicity should be respected.
- Information should be gathered and recorded systematically to ensure consistency.
- Straightforward, non-jargon language should be used.
- The issues of confidentiality and consent should be addressed.
- Assessment information gathered should be made available to the family.
- There should be clarity in respect of the lead role.
- Those undertaking needs assessments should have appropriate skills and knowledge.

In theory, assessment and care planning is simple:
- Perform a holistic needs assessment: itemising individual physical, psychological, social and spiritual needs of the child (and ideally for each member of the family too).
- Draw up and prioritise a problem list based on the needs assessment..
- Discuss and agree a management plan for each problem.
- Develop a multi-agency care plan.
- Involve the family.
- Agree a time for the next review.

In practice however, things can get more complicated

- Children's palliative care involves identification, assessment and management of a combination of physical, psychological, social and spiritual problems. Hence problem lists are often long and complex, requiring well organised, clear and up to date care plans.
- We tend to only 'see' those needs we feel confident and competent in addressing, and to 'ignore' those we don't. So a doctor's needs assessment often focus on medical problems and drug treatments, whereas those of a teacher will look at educational needs, and those of a chaplain spiritual needs. But of course a child has all these needs, so we need to 'think big' to do this well, even though it will mean thinking well outside our comfort zone.
- Decision making in children's palliative care is often difficult; involving discussions about prognosis, the risks and benefits of different therapies, the effects of complex family dynamics and the effect of ethical and legal issues.
- The question of how quality of life is best served is complex and consensus may be difficult to achieve.
- Problems such as guilt, fear, denial, avoidance and collusion from the child, family, or from professionals, can make the development of genuine problem lists and care plans very difficult.
- Needs of individual children are constantly changing with time, with their development, and with progression of their condition. These changing needs often require the input of many different professionals and voluntary carers, so there is potential for confusion and miscommunication.
- The needs of adolescents and young people are different from those of younger children and should be considered accordingly. Their emotional needs are likely to be more acute, and they will have additional issues such as body image, sexual needs, and a need for independence and life goals[5].
- The need for continuity of care dictates that comprehensive clinical records be kept and made available to all professionals involved in an individual case. This can pose significant challenges for over-stretched and under-resourced health management systems and services[6], working across different agencies (statutory and

voluntary) and settings (hospital, hospice, school, respite centre, home).
- Practically, it is near impossible to get everybody who cares for the child together at one time. If no one steps up to take the co-ordinating 'key worker' function, one of two things happen. Either nobody takes charge and the child and family are left in a vacuum, or everyone tries to take charge and numerous professionals overwhelm the family with numerous different assessments and management plans. What should be a smooth, integrated and reassuring process for the family can become a disjointed, fragmented and distressing one. This is why it is essential that a key worker or lead professional is identified as soon as possible following diagnosis.

Should I become a key worker/lead professional?

If you are asking yourself this question with an actual case, the answer is probably 'yes', because you have already identified both the need for coordination and the lack of a coordinator. At least, if you yourself can't do it, then you have responsibility for helping the child and family find someone who will.

A key worker or lead professional is responsible for co-ordinating the child's comprehensive care package, co-ordinating and liaising with all the professionals and services involved with the child, and being a main point of contact for the family. This does not mean you have to do everything, but it does mean that you have to try to ensure it gets done.

The key worker is often a community children's nurse or a professional within the multi-disciplinary community team. An ideal professional is one who has an ongoing and long term relationships with families, assessment and care management planning skills, used to networking and co-ordinating care for patients, sufficient influence with other agencies and professionals to be taken notice of, and premises and support staff that can be used to support the process. In practice, this is not always achievable, particularly in resource-poor settings, but the key thing is for someone to take the lead.

So why not you?

The assessment process and content

Doctors and nurses are used to exploring how illness affects a patient's whole physical, psychological, social and spiritual condition. However, in children's palliative care, you will probably need to be a bit more systematic, asking about things you may not be used to addressing.

Here are a few prompts to help you through the assessment:
- Explain the assessment process, signposting what will happen when, who will be involved and the purpose.
- Involve the child at an appropriate level.
- An excellent way to open the assessment process is to ask the child and family to talk you through in detail, exactly what happens in their 'typical day'. This will give you a wealth of information and is a good way of helping the child and family build confidence, overcome shyness and develop trust in you.
- Once trust is built, ask the family to talk you through their ideas, concerns and expectations about the condition and prognosis. This will help you understand and prioritise their problems.
- Have a look at *the Together for Short Lives Care Pathway*[7] to get an idea of how to systematically go through all aspects of the child and family's life, looking for issues that might be affecting their quality of life.

The assessment should cover the needs of the whole family, including fathers, siblings, grandparents and significant others identified by the child. The information that is gathered will include factual details of the child and family, details of the professionals and services involved with the family, medical information, functional abilities of the child, nursing and personal care needs, emotional needs, educational needs and the family's home circumstances.

The list below highlights the various different areas you should consider when performing a needs assessment.

Family
- Information – care choices
- Financial/benefits advice
- Emotional needs
- Siblings' well-being
- Family functioning
- Short breaks
- Quality of life
- Interpreter
- Spiritual, cultural and religious needs
- Transition to adult services
- Genetic counselling
- Contact details for professionals

Child or young person
- Symptom and pain management
- Personal care needs
- Therapies
- Emotional needs
- Information needs
- Respite
- Education
- Leisure/play
- Quality of life
- Nursing support
- Spiritual, cultural and religious needs
- Transition to adult services
- Independent living needs
- Follow- up (routine/emergency)

Environment
- Place of care
- Adaptations

- Risk assessment
- Home assessment
- Equipment needs
- Transport needs
- School/college/university

What other resources might I find helpful?

- Together for Short Lives (2003) *Guide to the Assessment of Children with Life-limiting Conditions and their Families.* Bristol: Together for Short Lives
- Together for Short Lives, 2009. A *Guide to the Development of Children's Palliative Care Services,* 3rd Edition. Bristol: Together for Short Lives.
- Together for Short Lives (2004) *A Framework for the Development of Integrated Multi-agency Care Pathways for Children with Life-threatening and Life-limiting Conditions.* Bristol: Together for Short Lives
- Together for Short Lives (2009) *A Family Companion to the Together for Short Lives care pathway* [online]. Available from <http://www.act.org.uk/page.asp?section=118&sectionTitle=Family+companion+to+the+Together for Short Lives+care+pathway>
- Department of Health (2010) *National framework for Children and Young Peoples Continuing Care.* [online] available from <http://www.dh.gov.uk/en/Publicationsandstatistics/Publications/PublicationsPolicyAndGuidance/DH_114784
- Department for Education (2009) *Early identification, assessment of needs and intervention - The Common Assessment Framework (CAF) for children and young people: A guide for practitioners.* [online] available from <http://education.gov.uk/publications/standard/publicationDetail/Page1/IW91/0709>
- The Scottish Government (2009) *Getting it Right for Every Child in Scotland* [online] available from <http://www.scotland.gov.uk/Topics/People/Young-People/childrensservices/girfec/programme-overview/Q/editmode/on/forceupdate/on>

Part five

How do I manage children with HIV&AIDS in CPC?

What you probably know already

- HIV/AIDS is still a very significant cause of childhood death in the world
- Palliative care also has an important role to play in the relief of distressing symptoms
- Symptoms caused by HIV&AIDS can be managed successfully using the same principles as with symptoms due to other pathologies

What you may not know

- Even in the era of anti-retroviral therapy (ART's), palliative care remains a crucial part of HIV&AIDS care, because treatment sometimes fails, and more often is not available or affordable
- HIV&AIDS is a multi-system, multi-organ disease. HIV virus causes pathology in two ways: by suppressing the immune system and also by directly infecting and damaging organs and systems (e.g. HIV encephalopathy and neuropathy, HIV bowel disease, cardiomyopathy, nephropathy, lymphocytic interstitial pneumonitis (LIP) and chronic lung disease with cor pulmonale)
- Despite the high prevalence of pain in AIDS, pain in children with AIDS is likely to be under-diagnosed and under-treated [8,9,10,11,12].

- It is important to look for opportunistic infections as a cause of pain and symptoms in HIV positive children
- Symptoms may be as a result of side effects to ART's
- Children suffer mental, psychological, and social distress, have to act as sole caregivers for family members, drop out of or interrupt school and are at increased risk of abuse,
- AIDS doesn't just affect infected children. Millions of children are orphaned around the world, and even more are left in deepening poverty that results from sick and dying parents [13] [14]
- It is not necessary to be an HIV&AIDS expert to provide good Children's Palliative Care, but you do need to know about side effects and interactions of ART's, which can be significant in palliative care settings.

What are the particular CPC challenges facing children with HIV&AIDS?

Children with HIV&AIDS face all the same challenges as children with other life-limiting illnesses, but there are some particular challenges too:
- Multiple symptoms
- Inadequate knowledge and skills in children's palliative care and management of childhood HIV&AIDS among care providers.
- Lack of prioritisation by governments, health and education systems
- Symptoms and manifestations of the disease change rapidly with disease progression
- Variability of access to ART and anti-TB treatment
- High rates of ART treatment failure in children (often due to poor adherence)
- Significant morbidity due to side effects of ART treatment
- Families refusing to engage in HIV treatment services for a variety of social and cultural reasons
- Lack of resources for identification, monitoring, treatment, and follow up

- Severe infrastructural weaknesses limiting access to health facilities, drugs, support and other treatment
- Prolongation of the chronic disease phase resulting in more complex and expensive management
- Wide range of drug toxicities of ART
- Competing effectiveness and toxicity of different drugs leading to complex decisions about sustaining or withdrawing treatments
- Emergence of co-morbid conditions (e.g. hepatitis, chronic psychiatric illness)
- Development of immune reconstitution illnesses in children started on ART's - especially in late stages of HIV
- Stigmatisation of AIDS resulting in failure to engage with healthcare services
- Guilt associated with mother-to-child-transmission
- Possibility/likelihood that more family members are infected, sick, or dying
- Many family members affected resulting in multiple bereavement and overstretched carers
- Breadwinners in families are often the most affected by AIDS

How do I work out the prognosis in a child with HIV&AIDS?

With the advent of ART, prognostication in HIV&AIDS has become extremely unreliable, as children apparently on death's door can make dramatic recoveries. It requires a very good understanding of both the evidence and of the specifics of the individual child (his or her nature, history, investigations, previous management and so on). Even then, prognostication is little more than educated guesswork, but the guess is often crucial to a decision which literally has life and death consequences. To help you, here are some indicators of a poor prognosis in HIV&AIDS[15]. However, all of these factors may potentially be over-ridden in the setting of effective antiretroviral therapy

- Laboratory markers suggesting poor prognosis
    - CD4 + T-lymphocyte count < 25 cells / mm^3
    - Cd4 < 15%

- Serum albumin < 2.5 gm / dl
- Clinical conditions with worse prognosis. This would basically be stage 4 HIV disease.
  - CNS lymphoma
  - Progressive multifocal leukoencephalopathy (PML)
  - Cryptosporidiosis
  - Severe wasting
  - Visceral Kaposi's sarcoma
  - Toxoplasmosis
  - Severe cardiomyopathy
  - Chronic severe diarrhoea
  - Life-threatening malignancies
  - Advanced end-organ failure (e.g., liver failure, congestive heart failure, COPD, renal failure, chronic lung disease)

Ultimately, it is almost certain that you will be called upon by a child's family to give your opinion as to the child's likely prognosis, because it is very stressful and exhausting not to know when death is going to occur. This stress and exhaustion can be complicated by guilt and anxiety triggered by wishing that everything could be all over with. In the author's experience, as long as you explain that you cannot be certain, it is usually possible to talk in terms of hours, days, weeks or months, but not more specifically than that.

What are the most common symptoms in children with HIV&AIDS?

The most common symptoms in children with HIV/AIDS are [16] [17]
- Fever, sweats, or chills (51%)
- Diarrhoea (51%)
- Nausea or anorexia (50%)
- Numbness, tingling, or pain in hands/feet (49%)
- Headache (39%)
- Weight loss (37%)
- Vaginal discharge, pain, or irritation (36%)
- Sinus infection or pain (35%)

- Visual problems (32%)
- Cough or dyspnoea (30%)
- Other common symptoms include earache from otitis media, gingivitis, dental caries, dystonic pain secondary to encephalopathy, fatigue, depression, agitation and anxiety,
- Nausea and vomiting

What causes pain in HIV&AIDS?

Pain in HIV/AIDS can be caused by
- Opportunistic infections (e.g. headache with cryptococcal meningitis, visceral abdominal pain with disseminated mycobacterium avium complex)
- The effects of HIV itself (e.g. polyneuropathy and myelopathy)
- Effects of medications used to treat HIV and the associated opportunistic infections;
- Procedural pain due to repeated procedures such as venesection, tube feeding, lumbar punctures and so on

Common types and sites of pain are given in the list below [18]

Pain Type (% of patients affected)
- Somatic pain, (71%)
- Neuropathic pain, (46%)
- Visceral pain, (29%)
- Headache, (46%)

Pain Site (% of patients affected)
- Joint, (47%)
- Polyneuropathy, (42%)
- Muscle, (40%)
- Skin, (23%)
- Bone, (31%)
- Abdominal, (25%)
- Chest, (19%)
- Radiculopathy, (18%)

How do I manage antiretroviral therapy in Children's Palliative Care?

It is very important to understand that significant drug interactions can occur in children receiving palliative care drugs who are also on ART's. Furthermore most of these medications may need to be administered in the presence of other co-morbid conditions such as hepatitis, pancreatitis, gastritis, hypertriglyceridemia, hyperglycemia, lipodystrophies, HIV-associated nephropathies and opportunistic infections. These can increase the risk of and the effects of interactions and adverse effects of drugs. It is beyond the boundaries of this book to deal with the whole pharmacology of ART's so please look elsewhere for this.

However, it is important to understand the Cytochrome P450 (CYP) enzyme system which metabolises most medications
- Changes in the CYP system can increase or reduce the amounts of circulating drugs, thereby affecting efficacy and toxicity
- Fortunately, the majority of drug-drug interactions are minor in nature and do not require extensive changes to the child's drug regimen
- Drugs that *inhibit* the CYP system cause the most dangerous interactions as they increase the blood levels of toxic drugs

Which drugs are most likely to affect the efficacy or toxicity of ARTs?

Certain drugs commonly used in CPC induce the CYP system and so may lead to low levels of ARTs, thereby increase the risk of drug resistance and/or treatment failure. These include
- Carbamazepine (Tegretol)
- Rifampin (Rifadin)
- Phenobarbital
- Phenytoin
- Prednisolone
- Cigarette smoke

- Omeprazole
- Isoniazid

Other drugs commonly used in CPC inhibit the CYP system and so may lead to increase ART levels and thereby increase the risk of toxicity. These include:
- Ketoconazole
- Itraconazole
- Erythromycin
- Fluoxetine
- Diltiazem
- Verapamil
- Clarithromycin
- Omeprazole
- Ciprofloxacin
- Fluconazole
- Metronidazole
- Trimethoprim/sulfamethoxazole (Septrin)
- Haloperidol
- Cimetidine

What are the highest risk drugs when used with ARTs which inhibit the CYP?

Certain drugs that are commonly used in CPC may reach toxic levels when given with ARTs which inhibit the CYP system. These include
- Tricyclic antidepressants (e.g. amitriptyline): risk of prolonged QT interval and sudden cardiac deaths
- Macrolides (e.g. erythromycin): risk of prolonged QT interval and sudden cardiac deaths
- Newer antihistamines (e.g. terfenidine): risk of prolonged QT interval and sudden cardiac deaths
- Cisapride: risk of prolonged QT interval and sudden infant death syndrome
- Quinine and chloroquine: risk of prolonged QT interval and sudden cardiac deaths

- Chloral hydrate: risk of prolonged sedation and respiratory depression
- Benzodiazepines: risk of prolonged sedation and respiratory depression..
- Methadone: risk of prolonged sedation and respiratory depression
- Rifabutin (Mycobutin): Ritonavir increases the risk of rifabutin-induced haematological toxicity by decreasing its metabolism.
- Clotrimoxazole/Sulfamethoxazole (Septrin): risk of increase in allergic reactions, especially rash
- Beta blockers: risk of significant falls in blood pressure and heart rate
- Haloperidol: risk of increased dystonic side effects and drowsiness

Even though these drug cross-reactions exist, it does not necessarily mean that the drugs listed above should be avoided. Indeed they may be very useful and helpful in individual cases. It is more that doctors and nurses need to be aware of the potential increased risk related to combined use, and carefully balance these against potential benefits. These risks and benefits should be clearly explained to the family, so that they can make an informed choice about them.

How do I counsel children and families about potential fatal interactions?

While children are generally less prone to cardiotoxicity than adults, this is not always the case, particularly where there are co-morbid cardiac conditions. All children using these drug combinations should be counselled to immediately report tachycardia, light-headedness, palpitations, vomiting or diarrhoea and avoid use of street drugs, substances of abuse, or excessive use of alcohol.

Practical management of symptoms in children with HIV&AIDS [19]
Individual symptom management advice is covered more fully in the relevant chapters of this book. However, to demonstrate the overlap between disease specific treatment and palliative treatment that is a feature of HIV&AIDS, the following table will give an overview.

Symptom	Causes	Disease specific therapy	Palliative therapy
Fatigue, weight loss, anorexia	HIV infection Opportunistic infections Malignancy Anaemia	ART Treat infections Transfusions Nutritional support	Explanation and reassurance Lifestyle modifications Steroids
Pain	See above	ART Treat specific diseases using antibacterials/ antifungals/ antivirals	Treat underlying cause Remember non-pharmacological approaches Consider ART Use WHO pain ladder
Nausea and vomiting	Drugs, gastrointestinal, infections	Stop drugs Treat infections using antifungals, antiparasitics, antivirals and antibiotics	Antiemetics Prokinetics H2 blockers (e.g. rantidine) or PPI (e.g. omeprazole). Small frequent feeds, fluids between meals, offer cold foods, eat before taking medications, dry foods, avoid sweet, fatty salty, or spicy foods. Aromatherapy, chewing ginger and altering feed regime in children on enteral nutrition.

Dysphagia	Candidal oesphagitis	Antifungals	If severe, reduce inflammation by giving steroids initially (may need IV initially). Use local guidelines or fluconazole or itraconazole. In resource poor settings we have had some success using clotrimazole pessaries 500mgs to be sucked daily for 5 days. Use analgesic ladder for pain
Sore mouth	Herpes simplex, aphthous ulcers, thrush, gingivitis	Acyclovir	Keep mouth clean; clean with soft cloth or gauze in clean salt water. Give clear water after each feed. Avoid acidic drinks and hot food. Give sour milk or porridge, soft and mashed. Ice cubes may help; ice cream or yoghurt, if available and affordable.
Chronic diarrhoea	Infections (gastroenteritis, parasites, MAC, cryptosporidium, CMV) malabsorption, malignancies, drug-related	Antibiotics/ antivirals/ anti-parasitics	Rehydration (see chapter), Vit A and Zinc, Diet modification, micronutrient supplements. Kaolin (cosmetic only) or bismuth. Oral morphine can alleviate intractable diarrhoea as can loperamide if available.

Constipation	Dehydration Tumours Drugs	Rehydrate Treat tumours with DXT or chemo if appropriate Adjust medication	Activity Diet modification Laxatives
Ano-genital ulceration:	Commonly due to herpes simplex virus Candidiasis	Herpes: Acyclovir (oral) or an emulsion mixture of Nystatin 5 mls, metronidazole powder 400mgs and Acyclovir 1 tablet Antifungals	Topical steroids may give temporary relief while the acyclovir takes effect Simple analgesics and adjuvant drugs for pain
Breathlessness	Pneumonia Anaemia Tumour Effusion Weakened respiratory muscles	Treat cause Antibiotics Iron or transfusion if severe Treatment of tumour (if appropriate) Drainage (if appropriate)	Fan and maximize airflow Counselling Distraction Relaxation Guided imagery Opioids Benzodiazepines
Persistent cough	Infections, LIP, bronchiectasis, TB, effusion, tumour	Antibiotics PCP treatment Anti-TB treatment Treatment of tumour (if appropriate) Drainage	Nebulization with physiotherapy Suppressant (e.g. low-dose morphine) Physiotherapy Humidification Steroids (LIP)

Severe dermatitis	Seborrhoeic dermatitis Infestations Folliculitis Fungal infection Hypersensitivity Renal and liver disease	Antibacterials/ antifungal/ anti-parasitics Hydration Steroids	Emollients, antihistamines, antiseptics, topical steroids. Antimuscurinic antidepressants (e.g. amitriptyline) Anxiolytics Keep nails short to minimize trauma and secondary infection from scratching
Shingles and post-herpetic neuralgia	Herpes Zoster	Aciclovir if caught early	Post herpetic neuralgia- use amitriptyline, valproate, gabapentin, phenytoin pregabalin or carbamazepine for shooting pain (but beware interactions with ARTs). Add morphine if necessary In resource poor areas sap from the frangipani tree applied to the vesicles (before they break) anecdotally helps
Convulsions	Infections and infestations, encephalopathy, malignancies, PMLE,	Phenytoin Phenobarbitone Carbamazepine	Diazepam/midazolam, phenobarbitone or paraldehyde for acute control, then convert to longer term therapy such as carbamazepine, phenytoin etc. Beware interactions between anticonvulsants and ART's

Metabolic disorders	Anticonvulsants, dextrose, mannitol, steroids		Rehydrate. Ensure good oxygenation. Give high energy, low protein feeds until disorder resolves. Treat individual cause
Fevers, sweats	HIV MAC CMV Lymphoma	ART Azithromycin Acyclovir Chemotherapy	NSAIDS Steroids Hyoscine Cimetidine
Pressure sores	Malnutrition Reduced mobility	Nutrition Mobilisation	See chapter Wound dressing, metronidazole powder to control odour, honey applications on clean, debridement if necessary
Delirium/agitation	Electrolytes disturbances Toxoplasmosis Cryptococcal meningitis IC sepsis	Correct imbalances and Rehydrate Antifungals and antibiotics	Assist orientation Haloperidol or promazine Benzodiazepines
Depression	Reactive Chronic illness	Play therapy Counselling Distraction Antidepressants	Counselling Distraction

What other resources may I find helpful?

- PEPFAR HIV/AIDS Palliative Care Guidance for the United States Government in–Country Staff and Implementing Partners
- http://www.state.gov/documents/organization/64416.pdf
- Amery, J. (Ed.) (2009) Children's Palliative Care in Africa. Oxford: Oxford University Press. Available from <http://www.icpcn.org.uk/core/core_picker/download.asp?id=204>

- FHSSA: A Clinical Guide to Supportive and Palliative Care for HIV/AIDS in Sub-Saharan Africa http://www.fhssa.org/i4a/pages/index.cfm?pageid=3489
- WHO 2012 WHO guidelines on the pharmacological treatment of persisting pain in children with medical illness.

Part six

How do I manage symptoms in children's palliative care?

How do I assess pain and distress in children?

What you probably know already

- Pain is a subjective experience, not an concrete entity. It can vary from person to person and situation to situation.
- Pain is hard to quantify, except by the individual affected by it, and even then only relative to his/her own experience of it.
- Without a good pain evaluation, it is difficult to treat effectively and promptly.

What you might find useful

How do I assess pain and distress?
You need to think about the possibility of pain being present in any palliative situation. Look for symptoms that are known to be painful or that can cause anxiety and pain. Remember that there may well be more than one pain and each one needs to be identified, quantified and treated. Because communication about pain is tricky, always try and support any information with as much evidence and common sense as possible. Also bear in mind that children may use the word pain to describe discomfort caused by other symptoms, such as nausea or dyspnoea. Therefore, a pain

assessment should always be part of a comprehensive review of systems or general symptom assessment.

There are three main ways to assess a child's pain:
1. **Ask the child**: The quickest and most accurate, providing the child is able to tell you (and you are able to understand).
2. **Ask the family (or known carer)**: The next best, and should be done as a cross-check even when the child has already told you, in case they are hiding pain.
3. **Try to assess it yourself**: The least accurate option, but better than nothing if you are stuck.

Asking the child
- Spend time establishing trust, listen carefully and don't underplay what you hear.
- Beware of asking leading questions – children may want to please you.
- Remember children may not realise how much they are hurting because they have been in constant pain[20] for some time, and parents may get desensitised to their child's pain for the same reason, and therefore fail to report it.
- The usual 'pain sieve' is just as useful in children's palliative care as in adults, but the information can be harder to obtain from children, and they may struggle to understand you. They may also struggle to understand you depending on their cultural context and developmental level.

Using the pain sieve
Below are some ideas of phrases and tips to use when implementing the pain sieve with children:
- *"Does it hurt or is it sore anywhere?"*
- *"Can you show me where it hurts/is sore?"* Use a body chart or a doll here.
- *"Does it hurt/is it sore anywhere else?"*
- *"When did the hurt/pain start?"*
- *"Does anything make it worse?"*

Children may have difficulty pinning down time frames until seven or eight years old, so you have to be a bit fluid in interpreting this. It is often useful to use a pain diary that the child or parent can fill in, so you can check progression, patterns and response to treatment.

- *"Do you know what might have started the hurt/pain?"*
- *"How much does it hurt/how bad is the pain?"*: Use a pain scale[21]
- *"Can you tell me any words that might describe the hurt/pain?"*
- *"What helps to take away the hurt/pain?" "What makes it worse?"*

Asking children to draw a picture to describe their pain or to colour on a diagram where their pain is at its worst can be very helpful (for example the Eland Colour Scale[22]

Asking the family or carer:
Try to find out what the parents' concerns are, and what they think are the reasons behind any behavioural changes. Remember parents may under report through desensitisation or fear. Ask the family or carer the same questions you asked the child, and you can use the same rating scales and charts with the parents as you did with the child. Ask the family or carers if they have noticed any signs of pain (particularly facial expression, body movements, quality of cry, stillness or withdrawal, change in favourite activities or sleeping etc.). Considering the impact of pain on the child's quality of life is also important (e.g. school attendance, sport, hobbies, mood, worry and sleep).

Trying to assess the pain in children who cannot communicate:

Remember, it's very difficult to assess pain objectively in children. Furthermore, as children get older or have a longer experience of pain, their facial expression and cry become less useful as indicators of pain. This may be because children in chronic pain 'down-regulate' normal pain behaviours and just become withdrawn. Also older children develop a much wider repertoire of behaviour learnt from those around them (including language).

However, assessment is still useful, particularly if a framework is used. Remember that pain generates a sympathetic nervous response, and therefore we can expect to see some of the following in a child who is in pain, although again the sympathetic response (and therefore clinical signs) may fade with chronicity of pain:

- Physiological: Increases in heart rate and breathing rate (but remember these can be drug induced or drug inhibited), pallor, sweating
- Behavioural: Quiet or crying, withdrawn or clingy, wincing or facial grimacing, moanin, restlessness, holding limbs, hands, feet or digits clenched or bunched

These are incorporated into the FLACC[23] (face, legs, activity, crying and consolability) scale.

In summary, you can pull all this together by using the QUEST approach:

Q: Question the child
U: Use pain rating tools
E: Evaluate behavioural cues
S: Sensitise the parents to be aware of and to report pain
T: Take action

Answers checklist for pain assessment

By the end of your pain assessment you should have answers to the following questions

- What words do the child and family use for pain?
- What verbal and behavioural cues does the child use to express pain?
- What do the parents and/or caregivers do when the child has pain?
- What do the parents and/or caregivers not do when the child has pain?
- What works best in relieving the pain?

- Where is the pain and what are the characteristics (site, severity, character of pain as described by the child/parent, e.g. sharp, burning, aching, stabbing, shooting, throbbing)?
- How did the present pain start (was it sudden/gradual)?
- How long has the pain been present (duration since onset)?
- Where is the pain (single/multiple sites)?
- Is the pain disturbing the child's sleep/emotional state?
- Is the pain restricting the child's ability to perform normal physical activities (sit, stand, walk, run)?
- Is the pain restricting the child's ability/willingness to interact with others, and ability to play.

(Adapted from the WHO guide, 2013)

How do I manage pain in children?

What you probably know already

- Chronic or severe pain is a common presenting problem in children with complex, life-limiting medical conditions, but less so in children.
- Pain is a multi-dimensional experience: influenced and caused by physical, psychological, social as spiritual causes (total pain).
- Physical pain is moderated through nerve pathways - from peripheral nociceptors, through the spinothalamic tract and the thalamus. Ascending pain signals can be reduced by 'descending inhibition' (e.g. due to relaxation, distraction etc.).
- Different drugs work at different points on the pain pathway.
- Choice and combinations of analgesics therefore can be rationally planned.
- The World Health Organisation Guidelines 2012 systematises choice and combinations of analgesia in all types of 'persisting' pain in children. It recognises the separation of mild vs. moderate to severe pain and the early and appropriate use of both simple and opioid analgesia.
- Other 'adjuvant' drugs can be used for certain types of pain, such as neuropathic pain or capsular distension pain.

What you might find useful

- Children in the CPC setting will not necessarily tell you they are in pain, even if they are, and they very rarely exaggerate it. So if they tell you they are in pain, it is best to believe them.
- In children, common factors affecting pain experience include fear, anxiety, separation from family, being in strange environments, co-existing illness/symptoms, and stigma.
- 'Total pain' is a term coined to express that pain is a crisis at every level[24]. It therefore follows that good pain relief entails managing all contributory physical, psycho-social and spiritual factors.

- Non-pharmacological pain management in children is proven in effectiveness and should always be included in any pain management strategy. Medications alone are never enough.
- Management of pain in palliative care is multi-modal and must be directed to all the possible areas by which pain can be modified. Analgesics are only one way of managing chronic pain. The list below outlines some other therapies that may be useful for pain management in paediatric palliative care.
    o Non-pharmacological management (see below).
    o Modification of the pathological process: e.g. by radiation therapy, antibiotics, hormonal therapy, chemotherapy, surgery and bisphosphonates.
    o Use of analgesics and adjuvants (see below).
    o Interruption to pain pathways: local anaesthetics, neurolysis and neurosurgery.
- 'Opio-phobia' is common amongst doctors and nurses around the world.
- The WHO no longer recommends codeine or dihydrocodeine for children, as most cannot metabolise them into effective metabolites. If paracetamol and non-steroidals are not enough, add oral morphine.
- There are more myths about morphine than any other drug. For the record, morphine used for pain does not cause addiction, morphine does not hasten death, and starting morphine does not imply you are giving up, nor does it mean that it is the 'beginning of the end'. The only thing that morphine will end is pain.

Non-pharmacological approaches to pain treatment in children

Non-pharmacological approaches can be highly effective, particularly in children, who are favourably suggestible[25] Non-pharmacological approaches should *always* be used in children's palliative care, as they help children cope with, and understand their pain and reduce anticipatory pain.

The largest effect sizes for treatment improvement over control conditions exist for distraction, hypnosis, self-reported distress, behavioural measures of distress, and combined cognitive-behavioural interventions. [26].

Non-pharmacological approaches are easy to learn and should be used in ALL situations, even where strong analgesia is used to be used in addition.

Some examples of non-pharmacological approaches include:
- Reinforce coping behaviour by reminding and congratulating the child about how well they have coped (even if they have not coped brilliantly). Do not ignore problems that happened, but talk them through [27] and look at them from a more positive light in order to try and reduce fear of them.
- Distraction. (link to distraction techniques table)
- Deep breathing[28]
- Progressive Muscle Relaxation[29]
- Hypnosis
- Guided Imagery. Children are experts at this, as they find it easy to use their imagination to take them away to a pleasant comforting place.
- Accupuncture/Accupressure
- Reiki
- Massage
- Aromatherapy

Principles of analgesia usage in children

The principles of analgesia in palliative care have been encapsulated by the World Health Organization (WHO)[30] in three slogans:
- By the mouth – avoid injections where possible, and if oral dosing is not possible consider rectal or transdermal.
- By the clock – persistent pain requires preventive, regular treatment.
- By the child – use the 2 step WHO analgesic strategy (Step 1: Simple analgesia. Step 2: opioid, with or without adjuvants if necessary).

How to use opioids

There is considerable confusion about how to use morphine effectively in CPC. To understand it properly, we need to remember two key facts. These are that, in a child with relatively normal intestinal, liver and renal function:
1. The average time to maximum plasma concentration (Cmax) after a dose of oral morphine is ONE hour
2. The average duration of action of oral morphine is THREE TO FOUR hours in children older than around six months

If Cmax is one hour, then it follows that, if pain is uncontrolled after one hour, analgesia is not going to get any better, so the 'AS REQUIRED' dose should be given every ONE hour (not every two, three or four hours as most formularies have it).

If the duration of action of morphine is four hours, it follows that the REGULAR dose of oral morphine (once good pain control is achieved) should be given every FOUR hours.

The 'total daily dose' (TDD) is the sum of all PRN and regular doses given in the last 24 hours

As there are six doses in 24 hours, the next day's 'regular' and 'as required' dose should be ONE SIXTH of the previous 24-hour's total daily dose.

Keep using one-hourly PRN dosing ALONGSIDE 4-hourly regular dosing until either the pain is well controlled (no more than one PRN dose required in 24 hours) or the child begins to exhibit significant side effects such as pinpoint pupils, twitching, urinary retention and so on.

If pain relief is not achieved despite hitting significant side effects, consider rotating to a different opioid (such as oxycodone) or using adjuvants.

Opioid toxicity in children

Children usually tolerate opioids quite well, but there are some common side effects.

- Constipation is common and children should be prescribed a laxative early on when opioids are started to prevent painful large stool (which can lead to retention and impaction of stool). It is important to recognise the need for a stimulant laxative or combination. It should not be just a stool softener such as lactulose. Lactulose is one of the most comment laxative in non-palliative care paediatric practice and therefore the most familiar but it is recognised that often children are given inadequate laxative prophylaxis in palliative care because there is a lack of understanding of the pharmacological effects of different laxatives.
- Nausea and vomiting is less common, and tends to wear off after a few days, but some children need antiemetics.
- Drowsiness is quite common in the first few days but this also improves with time.
- Itching and urinary retention are not uncommon and, if they occur, you may need to try alternative opioids.
- Significant toxicity are relatively infrequent, and less likely to occur if doses are titrated as outlined above. You may notice slowed breathing, pupil constriction and a reduced conscious level. These symptoms may indicate the need for an opioid-reversing agent, Naloxone, although the author has never had to use this in 20 years of CPC practice. Children can also develop restlessness, malaise, disorientation, excitation, agitation, twitches and jerks at higher doses of opioids, which may confuse us into thinking that the child is still in pain. These symptoms may indicate benefit in rotating to another opioid, such as oxycodone.

Using other strong opioids

There are a number of other opioids, with different preparations using different routes of administration. We have not included tramadol and methadone here as, although they have been used in the CPC setting, evidence for their use is more limited and methadone is quite complex to start without specialist advice.

- **Oxycodone is probably the most useful 'second line' opioid where a child has not responded to morphine or had inhibiting adverse effects.** There is no guarantee that oxycodone will not create the same side effects but it is often worth trying.
- **Diamorphine** is more soluble and also has better compatibilities for mixing with other drugs in a syringe driver. It may also be used nasally or buccally. (This is particularly useful in an emergency and for acute breakthrough pain rather than resorting to injections). However, this is an unlicensed route and is similar to an intravenous bolus dose in both potency and duration of effect, so take the same caution you would for an IV dose.
- **Buprenorphine** has a longer duration of action than morphine (six to eight hours) and can be used sublingually and transdermally. It does not rely on renal excretion and therefore can be used when renal failure is a problem. It has opioid agonist and antagonist properties and may precipitate withdrawal symptoms, including pain, in children dependent on other opioids. Buprenorphine is also now available as 7 day patches which can be useful in managing stable opioid-responsive pain especially at low total daily doses.
- **Fentanyl**[31] has been developed as 72 hours transdermal patches, which can be useful in a child with poor oral intake but who is too active for SC infusions to be practical. Although metabolised in the liver, its metabolism seems to require very little liver function so it is much less likely to accumulate compared to other opioids in children with hepatic failure. If using patches, you need to know whether the patch you prescribe is either reservoir or matrix patch as efficacy may vary if the preparation is changed. Fentanyl is also available as a nasal spray and as buccal or sublingual tablets

or 'lollipops' (lozenges): these can be useful as quick acting preparations for breakthrough symptoms and incident pain.

NB Caution with fentanyl and buprenorphine patches: if the underlying skin is vasodilated for example due to pyrexia or external warming, absorption of the drug can be more rapid, leading to unexpected toxicity.

Avoid Codeine

Codeine is not recommended for use in CPC. It is a 'pro-drug' which is converted in the body to morphine using an enzyme called CYP2D6, which is largely absent in the foetus and even at the age of 5 may be present at only 25% of adult levels. Its activity in some ethnic groups is also reduced.

Oral analgesic equivalence to morphine[32]**:**

Analgesic -Route-Dose
(PO = by mouth; IM = intramuscular, IV = intravenous, SC= subcutaneous)
- Diamorphine -IM, IV, SC-3 mg
- Dihydrocodeine - PO - 100 mg
- Hydromorphone - PO - 2 mg
- Morphine - PO - 10 mg
- Morphine - IM, IV, SC -5 mg
- Oxycodone - PO - 6.6 mg
- Tramadol - PO - 100 mg

Particular Pain Situations

Cancer Related Pain
Cancer pain is very common and can be due to a number of causes: invasion of local tissues, localised pressure effects, organ distension, regional inflammation, bowel or other luminal obstructions, and nerve

irritation. It is common also with invasive procedures and also as a side effect of chemotherapy.

Cancer related pain should be treated initially using the standard WHO approach described above. However, there are other options for cancer pain management, depending on the site and type of the cancer. For example
- Radiotherapy for localised disease
- Palliative chemotherapy
- Local nerve blocks
- Intrathecal and epidural infusions
- Palliative surgical resections

Where available, it is therefore important to liaise closely with local oncology and anaesthesic services, and make use of them wherever possible.

Neuropathic/nerve pain:
Suspect neuropathic pain if the child describes the pain using words such as 'burning', 'shooting', 'stabbing'; or if there are other neurological features such as sensory changes in the area of the pain, or if the pain is distributed in a dermatomal fashion.

There is a myth that neuropathic pain is insensitive to standard analgesics. This is untrue. They can be helpful, but may be only partially so. However, it is always worth trying the standard analgesic approach outlined above.

If the pain does not respond fully to this, try an adjuvant medication. These are drugs that stabilise nerve cell membranes, thereby depressing nerve activity. Not surprisingly, they usually suppress general activity too, and can be quite sedating and constipating so need careful titration. Evidence for these are limited, but amitriptyline and gabapentin are widely used and safe in children, freely available and cheap, so these should be first line. Generally choice is determined by compliance issues. Gabapentin needs to be taken three times a day) and whether additional properties of the medication might be helpful (for example amitriptyline has a sedative effect and may improve disturbed sleep). Other tri-cyclics anti-depressants and anti-epileptics such as carbamazepine, and pregabalin can also be tried. In particularly resistant neuropathic pain a tricyclic and anti-epileptic drug are used in combination.

Chronic non responsive pain:
If pain does not respond to analgesia, or if side effects become too troublesome, the first step is to try to rotate to a different opioid, for example from morphine to oxycodone, fentanyl (or alfentanil), or buprenorphine. Methadone is anecdotally better at reducing the increased sensitivity to pain and reduced sensitivity to opioids that occurs sometimes when children are on long term opioids (believed to be due to changes to central neurotransmitters and receptors, especially NMDA). It also has effectiveness for neuropathic pain. However, it has complicated pharmacodynamics it is probably wise to seek specialist advice before attempting to start it.

Raised intracranial pressure (and tumour compression in other areas):
Where pain results from compression of a closed site (for example the brain in the skull, or liver in its capsule) corticosteroids (especially dexamethasone) reduce inflammation and swelling in tumours very effectively. Long term use is NOT advised in this situation as side effects in children, particularly behavioural, emotional effects and the development of cushingoid features, can be more severe than in adults. Short pulses of high dose steroids can be helpful.

Spasticity/ increased muscle tone:
Antispasmodics such as baclofen and benzodiazepines such as diazepam and lorazepam may act as muscle relaxants but also have sedative effects – see the section on spasticity and dystonia.

Bone pain/soft tissue pain:
Non-Steroidal Anti-Inflammatory Drugs (NSAIDs) are often very effective in bone pain. Beware of using NSAIDs in any situation where there is bone marrow failure, as bleeding is much more likely. Advise to take after food and cover with a proton-pump inhibitor if available, or ranitidine if not as gastritis can be problematic.

Colicky abdominal pain:
Anticholinergics such as hyoscine butylbromide or mebeverine may be a helpful adjuvant medication for colicky pain, but can contribute to constipation.

Anxiety:
Try addressing issues and using non-pharmacological measures and if they are not effective, consider using anxiolytics such as low dose lorazepam.

A practical approach to pain management in children

While it is helpful to know the theory of pain management in CPC, it is arguably more helpful to know what to do in practice. When confronted with a child in pain, one's memory can become a bit hazy just at the time you need it to be slick and sharp. Half-remembered and unwanted physiological and pathological facts can clutter your mind just at the wrong time. In this situation, 'recipe book medicine' can be just what the doctor ordered.

Is the child in pain?
- Assess carefully to ensure you are dealing with pain, not other distressing symptoms such as anxiety, nausea etc.

Is the pain severe?
- If so, before you go any further, **relieve the pain**. Further assessment can wait until the child is comfortable. Give a dose of analgesia immediately.

What are the likely causes?
- Remember the child can have several causes of pain simultaneously. Assess and visualise the pathology until you have worked out what the cause of the pain(s) might be: e.g. compression, infiltration, infection, drug side-effect, neuropathic, psychological, and spiritual. Use of a body chart and pain scores can help clarify and monitor.

Is there a remediable cause?
- If there is, remedy it. If the bed is wet, change it. If there is an infection, treat it. If the bone is broken, immobilise it. If the child is constipated, ease it.

Is the child properly hydrated, nourished and oxygenated?
- Imagine how much worse pain is when you are cold, tired and hungry. Good nutrition, hydration and tissue oxygenation can avoid further stress in a painful situation.

Is the child comfortable, comforted and distracted?
- Use warmth, swaddling, feeding and reassurance. Handle gently and use supportive positioning to minimize pain from movement. Minimise invasive procedures. Use distraction, massage, relaxation and imagery.

Is the pain related to movement?
- This suggests musculoskeletal cause. Think of fractures or sprains (immobilise), metastasis (orthopaedic surgery or radiotherapy), soft tissue infection (especially myositis in AIDS - use antibiotics, drainage and immobilization), nerve compression, spasticity (can be very painful, see section on dystonia).

Is the pain due to organ distension?
- In organs which cannot easily expand (e.g. brain, liver) think about the use of radiotherapy or steroids (but take care using steroids in children – see above) or bladder – put in a catheter.

Is the pain due to cancer?
- Consider whether radiotherapy, chemotherapy, nerve blocks or surgical procedures may be helpful

Are there neuropathic features?
- If the pain is shooting/burning, is dermatomal in distribution and/or is associated with neurological signs, try standard analgesics but you may need adjuvants, e.g. amitriptyline.

Is there pain due to muscle spasm?
- Use warmth, gentle stretching, massage, splinting and physiotherapy. Try benzodiazepine or baclofen.

Controlling severe pain at the end of life

Pain control at the end of life follows the same general principles as presented in this section. If you need to control acute, severe pain for the first time at or near the end of life, and you do not have parenteral access, you can use subcutaneous morphine or buccal or nasal diamorphine as above. Please see the chapter on managing acute, distressing terminal events for more information.

How do I manage fluids and electrolytes in children?

What you probably know already

- As children head towards death, their desire to eat and drink declines so dehydration towards the end of life is commonplace.
- If a patient is conscious it is easy to find out if they are thirsty, but not if they are unconscious or confused.
- The younger the child, the more vulnerable they are to water loss.
- Feeding and drinking are more than just medical needs. They have profound cultural symbolism, especially for parents, so stopping them is a decision fraught with psychological, social and ethical dilemmas.

What you might not know

- Evidence regarding the incidence, effects and management of dehydration in the palliative care situation is sparse, contradictory and therefore often unhelpful[33,34,35,36,37,38,39,40,41].
- Dehydration in normal (i.e. non-palliative) situations can be distressing and have unwanted effects but in the terminal phases, reduced fluid intake may have benefits too (e.g. reduced incontinence and pulmonary oedema/secretion, reduced peripheral oedema).
- Therefore, the decision about whether to use artificial hydration when a patient is not able to drink and complaining of thirst is a tricky one.
- There are no golden rules – treat each case on its own merits.

Assessment of dehydration

It is useful to assess the degree of hydration, partly to plan treatment, and partly to monitor response.

Signs of 5% dehydration
- Loss of body weight, dry mucous membranes, slightly reduced urine output

Signs of 10% dehydration
- Decreased turgor with poor capillary return, increased capillary filling time, dry mucous membranes, depressed fontanelle, sunken eyes, poor volume pulses, lethargic, oliguria

Signs of >10% dehydration
- Acidotic breathing, tachycardia, tachypnoea, shock, anuria, eventual coma

A practical approach to children who are not drinking

Is the child thirsty?
- If so, give fluids either orally if possible, or by SC, IV or by feeding tube.

Is the prognosis short and the child either comfortable or unconscious without agitation?
- It is unlikely that a comfortable or unconscious child is suffering as a result of not drinking, so there is no ethical reason to give fluids, but there may be strong psychosocial reasons. Take each case on its own merits.
- The family will need compassionate and patient explanation and support.[42]
- Treat with good mouth care.

Is the child uncomfortable, agitated or confused?
- Consider the ethics of non-oral routes: One the one hand, artificial hydration may reduce terminal agitation, but not always. On the

- other, it may also precipitate heart failure and cerebral oedema, making the agitation and discomfort worse not better.
- If the law of your country is unclear or designates artificial hydration as an artificial treatment, decide with the child, family and colleagues whether the benefits outweigh the burdens or vice versa (see the chapter on ethics for more details).
- A pragmatic approach, if the family insists that feeding and or hydration should continue, would be to artificially hydrate (as your refusal may risk complicating their bereavement, while SC or tube rehydration is unlikely to cause significant discomfort to the child)
- There are a number of options for fluid administration in the terminal phase of a child's life[43]

In the end of life scenario, a child's fluid requirements will only be around 60% or less of their standard age/weight requirements, so take care not to overload. Intravenous infusions are possible but are cumbersome. Hypodermoclysis (subcutaneous fluid infusion) is less cumbersome and easier to manage in the home setting. Fluid infusion rates can often be run up to the maintenance hourly rate that would be given intravenously, with close and careful continual reassessment of the subcutaneous site. Nasogastric tubes can be useful in the short term but ares probably too distressing to be considered an ethical end of life treatment. Proctoclysis (rectal administration) is used in some countries

Justin Amery

How do I manage feeding problems in children?

What you probably know already

- In resource-poor settings, malnutrition in CPC is usually simply because children do not have enough to eat.
- There are number of other reasons why children may not eat enough in a CPC setting, including simple anorexia, sore mouth, difficulty swallowing, nausea and vomiting or gastric compression.
- Whatever the cause, if a child is hungry, it is the child's basic human right to be fed.

What you may not know

- Malnutrition used to be a big problem in CPC, even in resource-rich settings, when children were unable to swallow for neurological or neuromuscular reasons.
- With the advent of permanent tube feeding malnutrition is less of a problem in resource-rich settings but continues to be a huge problem in resource poor settings[44][45]. Leaving a child hungry makes a mockery of everything else in this book, so it's crucial to *find food first*.
- Not all children who have stopped eating need to be given food, for example if they are not hungry or when artificial feeding may cause more harms than benefits.
- Tube feeding has revolutionised feeding for children who can't manage to eat, but it is not without significant side effects, including gastro-oesophageal reflux, recurrent aspiration and chest infections, discomfort from the tube, abdominal pain and diarrhoea

A practical approach to children who are not eating:

Is the child hungry?
- This is a crucial question: a child with cachexia will rarely be hungry, whereas a child with mild or moderate malnutrition will be. Take care though. Most children with severe malnutrition have lost their appetite.
- Use the WHO criteria of weight for height and height for age to determine degree of malnutrition. If the child is hungry and malnourished, help him or her feed and drink with whatever tools you have available (following the WHO Malnutrition guidelines[46])

Is the prognosis is in days rather than weeks and months? .
- Give food or drink for pleasure, but not if the child resists. Remember families may often be reluctant to withdraw feeding even well after they have agreed to withdraw all other interventions and treatments. You might (or probably will) feel under pressure from the parents or family to continue feeding.
- In the terminal phase use compassion, skill and patience to help the family understand that food is not needed by the body at this point because it won't prolong life and the child is not feeling hungry (so there is no question of the child suffering).
- Remind them that the natural bodily process of dying does not include feeding. (This is obviously easier for parents to understand if the child is not conscious). Often giving mouth care (moistening the mouth with water and protecting lips from dying out) helps parents to make the transition from actively wanting to feed. Explain to them that forced feeding will cause suffering.
- Manage treatable causes
- Manage sore mouth, swallowing problems, nausea, squashed stomach, breathlessness, pain, renal or liver failure, chronic infection.
- Think about disability (a child may need help with feeding and positioning to eat properly).

Are you dealing with drug side effects?
- Think of opioids, metronidazole, cotrimoxazole, NSAIDs, chemotherapy, some antibiotics.
- Change drugs if possible.

Are there problems with the food?
- Check that food being offered is known and liked by the child and is culturally acceptable.
- Try using low bulk and high energy/high protein foods (if available).
- Ensure that the child can sit up and has company around the bed.
- Avoid strong odours that may cause nausea. Also avoid overly spicy or fatty foods.
- Do not pressurise the child but tempt him or her with small, frequent helpings/ tasters of favourite foods on the smallest plate available.
- Be ready to permit and encourage any bizarre fancy the patient may have. [47]

Is appetite stimulation available or appropriate?
- You can try megestrol (in older children with cancer) or a short course of low dose dexamethasone particularly if the child is nearing the end of life.
- If still no better, treat symptomatically
- For feeding (or as an alternative to non-oral hydration) consider tube feeding (either nasogastric or gastrostomy if available).

How do I manage sore mouth in children?

What you probably know already

- Sore mouth is common in palliative care, for example due to cytotoxic and anticholinergic drugs, reduced oral intake, dehydration, debility and local lesions.
- Oral candidiasis is a very frequent cause of sore mouth in CPC, particularly with HIV/AIDS.
- Simple mouth care is effective: such as keeping the mouth moist, gentle tooth brushing, gargling, using ice lollies or pineapple (which contains an enzyme which helps clean the mouth).

What you may not know

- Children with sore mouths may present with difficulty in swallowing, vomiting, reluctance to take food, excessive salivation, or crying while feeding.
- Aphthous stomatitis and recurrent herpes simplex are common in HIV/AIDS as is oral/oesophageal candidiasis

A practical approach to the management of mouth pain

General treatment
- Paracetamol before meals, avoid spicy foods and use soothing liquids or ice chips/ lollies

Dental caries and other infections?
- General mouth care
- Amoxicillin and/or metronidazole for infections

Non-specific spongy red mucosa, angular stomatitis, and/or white spots visible?
- Probable oral candidiasis (present in 75% of patients)
- Consider nystatin drops or lozenges

- Daktarin oral jel
- Fluconazole orally
- Amphotericin

Cytotoxic therapy?
- Marked stomatitis common
- General mouth care with an alcohol free, anti-bacterial mouth wash
- Use soft bland foods

Shallow, painful, cratered lesions with a raised, erythematous border and a grey, central pseudomembrane, may be small or large?
- Probable aphthous stomatitis
- Steroid creams (in orabase if available) or steroid lozenges held against the lesions
- Prednisolone tabs crushed and the powder applied topically.

Painful ulcers, may start with itching/tingling sensation, followed by painful vesicles, may be both inside and outside the mouth?
- Probable herpes simplex
- Acyclovir (IV if immunocompromised)

How do I manage dysphagia in children?

What you probably know already

- Dysphagia is very common in the CPC setting.
- We see two different scenarios in CPC:
    - Children who are frightened that they might choke or inhale food, and so avoid eating, leading to potential malnutrition.
    - Children who have never been able to swallow normally and are fed orally despite choking episodes and recurrent episodes of aspiration
- Dysphagia has many possible causes: In the lumen (e.g. foreign body, stuck food); in the wall (e.g. mucosal inflammation, candida infection, drugs damage, gastro-oesophageal reflux and spasm, neurological disorder); and outside the wall (e.g. mediastinal lymphadenopathy/tumour).

What you may not know

- In resource-rich countries dysphagia is usually due to neurological or neuromuscular conditions.
- In resource-poor settings dysphagia is very common in children with the above conditions, as well as with HIV/AIDS and cancers.
- Candidiasis is very common in the CPC setting, particularly in resource-poor settings, and particularly in HIV/AIDS.
- Gastrointestinal dysmotility and reflux is common in children with neurological and neuro-muscular conditions.

How to assess a child with dysphagia

If the child has cancer or HIV/AIDS
- Strongly suspect candidiasis in all cases (particularly if the dysphagia is painful, and/or the child has had steroids or chemotherapy). The absence of oral candidiasis does not exclude

oesophageal candidiasis. Consider it likely if the child refuses feeds, is drooling (unable to swallow own secretions) and has a hoarse cry from laryngeal candidiasis.

If there is dental pain/trismus
- Think of dental caries and abscesses.

If the child can swallow fluid easily, but food gets stuck
- Think of a stricture from chronic reflux or external compression form mediastinal lymphadenopathy.

If the child struggles to swallow liquids
- Think of a motility disorder, a neurological disorder, or a severely narrowed lumen.

If the neck bulges or gurgles on drinking
- Think of a pharyngeal pouch (uncommon).

If there is cough and recurrent chest infection with weak gag reflex
- Think of recurrent aspiration.

A practical approach to the management of dysphagia

All children
- Treat the cause if possible
- Estimate feeding and hydration and manage as above
- Use feed thickeners for children with reflux.
- Use acid neutralising treatments such as PPIs or ranitidine.

Candidiasis
- Nystatin is the cheapest and most widely available drug but if the candidiasis is deep set you may need to use oral antifungals for 2-3 weeks (e.g. fluconazole)

Drug related
- Stop NSAIDs, steroids and anticholinergics if possible

Recurrent aspiration
- Consider tube/ PEG feeding.
- Discuss with family risks and benefits of continued small amounts of oral feeds for pleasure/quality of life.

Tumours
- Refer to oncology if available.
- Consider dexamethasone to reduce tumour bulk
- Consider stenting or gastrostomy if available

Pharmacological symptomatic treatment
- Consider acid suppression with PPI (e.g. omeprazole) or H2 blockers (e.g. ranitidine)
- Try a motility stimulant (e.g. domperidone, low dose erythromycin or metoclopramide where this is still licenced for use in children)
- Local anaesthetic (e.g. lidocaine) mixed with antacid formulation.
- Treat residual pain using the WHO approach

Justin Amery

How do I manage gastro-oesophageal reflux

What you probably know already

- Reflux is common and normal, to an extent. But it is abnormal if it causes symptoms or complications.
- There is a poor correlation between the severity of reflux and the symptoms it causes.
- It can be caused by raised intra-abdominal or intra-gastric pressure (e.g. from dystonia, tube feeding, scoliosis), faulty gastro-oesophageal sphincter (e.g. NG tubes, hiatus hernia, bowel dysmotility) and drugs (particularly anticholinergics, steroids and non-steroidal anti-inflammatory agents).
- It can be investigated by a therapeutic trial with appropriate medicines, endoscopy, 24 hour oesophageal pH monitoring, barium studies or videofluoroscopy (if aspiration or poor swallow are suspected).

What you may not know

- Reflux is very common in the CPC setting, because many children have bowel dysmotility, dystonia, are tube fed, or are on precipitating drugs.
- It often presents with atypical symptoms in children such as failure to thrive or weight loss, feeding refusal, arching or strange posturing (often confused with dystonia or seizures), disturbed sleep, wheezing, recurrent chest infections, hypersalivation, recurrent emesis or haematemesis.

A practical approach to the management of reflux

Diet
- If orally fed: Try thickening the feed using commercial thickeners or rice or cornflour. Give frequent small meals.
- If tube fed: Anecdotally, one of the major causes of reflux in CPC is tube feeding. Try adjusting the rate/volume, or frequency of feeds. Be cautious about relying entirely on nutritional charts for calculating input. We have seen some children actually *gain* weight after *reducing* feeds, presumably because of reduced intestinal hurry and better absorption.

Positioning
- In babies a prone position is better than a supine position (although not in sleep due to risk of sudden infant death syndrome).
- In older children lying on the left side and sleeping with the head of the bed elevated may help
- Stop causative drugs where possible
- If possible stop or reduce drugs, particularly anticholinergic, non-steroidal anti-inflammatories, steroids, benzodiazepines, some antibiotics.

Drugs treatment
- Try antacids and raft forming agents such as alginates (e.g. gaviscon).
- Choose between H2 antagonists (e.g. ranitidine) or proton-pump inhibitors (e.g. omeprazole or lansoprazole).
- Domperidone may improve gastric emptying as may low-dose erythromycin and metoclopramide.

Surgery may be indicated
- Fundoplication is often tried in resource-rich settings but there may be a high rate of failure[48].

How do I manage gastro-intestinal bleeding in children?

What you probably know already

- Bleeding of any sort is a very frightening event
- Bleeding can be due to oesophagitis, varices, gastritis, drugs, tumours, clotting disorders, swallowed blood from the nose and mouth, volvulus, Meckel's diverticulum, and arteriovenous malformation,[49]

What you might not know

- Massive life-threatening GI bleeding is not as common in Children's Palliative Care as in adults, but it does occur
- Low grade GI bleeding is common in CPC due to the high rates of reflux and gastritis

Practical Assessment of GI bleeding in children

Is the child vomiting frank blood or coffee grounds, or just some blood mixed with the vomit?
- Large volume frank red blood: think oesophageal causes
- Large volume "coffee grounds" +/- melena: think gastric bleeding, probably ulceration.
- Small amount blood staining in vomit: think swallowed blood (look for bleeding in nose, mouth and pharynx)

Does the child have retrosternal pain and/or dysphagia?
- Think oesophagitis.

Does the child have epigastric pain?
- Think of gastritis/duodenitis or ulceration.

Is there a known bleeding disorder, or signs of bleeding elsewhere?
- Think clotting abnormalities.

Is the child getting central colicky abdominal pain?
- Think small or large bowel causes such as Meckel's diverticulum, volvulus or AV malformation.

A practical approach to the management of GI bleeding

All children
- DON'T PANIC: Keep calm, reassure the child and family and give them something to do to distract them (e.g. 'fetch towels/bowls/music/tv etc) Position child so that blood/vomit drains easily.
- Provide receptacles and dark towels to collect blood.
- Consider siting an IV cannula for drugs and/or fluids/blood if appropriate and available.
- Check FBC and clotting screen if available.
- Consider referral for surgery/endoscopy (if available and appropriate).

Mouth bleeding
- Simple mouth care, tranexamic acid mouthwash, adrenaline soaked into gauze and applied directly or, if available. 'gelfoam' (gelatin granules on an absorbable gelatine sponge).
- Stop causative drugs if possible, for example NSAIDs, steroids, warfarin.

Clotting disorder
- Consider platelet infusion if platelets are low.
- Consider vitamin K if there is liver dysfunction.

Tumour
- Consider radiotherapy referral.
- Consider use of tranexamic acid (if available).

Massive life-threatening bleed?
- See the section on 'acute, distressing terminal events'.

How do I manage nausea and vomiting in children?

What you probably know already

- Nausea and vomiting are common presenting problems.
- Nausea experiences are affected by psychological and social as well as physical causes.
- Nausea and vomiting are modulated via various receptors in the gut, vestibular apparatus and the brain.
- Different drugs work at different receptors.
- Choice and combinations of anti-emetics can therefore be rationally planned.
- Other 'adjuvant' drugs can be used for some nausea and vomiting, such as nausea and vomiting due to anxiety.

What you might find useful

- Children vomit very easily, but not always for the same reasons as adults. The prime difference is that children, particularly small children, have very sensitive gag reflexes, therefore any irritation to the back of the throat can trigger vomiting.
- In our experience, nausea and vomiting in children is often (inadvertently) caused by drugs prescribed by health workers.
- Nausea is a subjective experience and, like pain, its intensity is influenced by emotional factors (e.g. anxiety, fear, isolation) and cognitive factors (e.g. meaning and memory of similar).
- It therefore follows that good pain relief entails managing all contributory physical, psycho-social and spiritual factors.
- Most health professionals now know that pain can be treated in a rational, step-wise approach, but fewer know that nausea and vomiting can also be treated in the same way

Principles of nausea and vomiting treatment in children

- The table below shows the sites of the various receptors involved in nausea and vomiting and the transmitters acting at them
- If you can bear with the pharmacology a little longer, you can see from the table below[50] which anti-emetics work on the different receptors.

DRUGS	EMETIC RECEPTOR ACTIVITY					
	Dopamine	Histamine	Muscarinic	Serotonin (5HT2)	Serotonin (5HT3)	Serotonin (5HT4)
Domperidone	++	0	0	0	0	0
Ondansetron	0	0	0	0	+++	0
Cyclizine	0	++	++	0	0	0
Hyoscine Hydrobromide	0	0	+++	0	0	0
Haloperidol	+++	0	0	0	0	0
Promazine	+++	+	0	0	0	0
Metoclopramide	++	0	0	0	+	+
Levomepromazine	+++	+++	++	+++	0	0

- If in doubt, probably the best first line anti-emetic in children is cyclizine
- Domperidone is useful where gastric emptying or GI dysmotility is a problem
- For raised ICP choose cyclizine and/or ondansetron, and consider augmenting with dexamethasone
- For drugs-related, metabolic, infective or biochemical upset: try haloperidol
- Where anxiety or agitation are factors, try haloperidol or levomepromazine
- The broadest spectrum antiemetic is levomepromazine, so it is very useful in CPC, but its use is limited by the fact it is very sedating (which is not necessarily a problem right at the end of life, but is when the child wishes to remain active).
- Ondansetron, often thought of as an antiemetic wonder drug, is actually quite limited in its receptor effects, but is very useful

in chemotherapy-induced vomiting because chemotherapy hits 5HT3 receptors, which are not well modulated by any other antiemetics.
- It is worth noting that cyclizine and metoclopramide should not be prescribed together as they have antagonistic properties
- Recent pharmacological warnings in Europe have advised against prolonged prescribing metoclopramide due to adverse neurological side effects[51] but it is still one of the few available anti-emetics in many countries, and these risks need to be balanced against the harms of allowing a child to suffer with untreated nausea and vomiting.

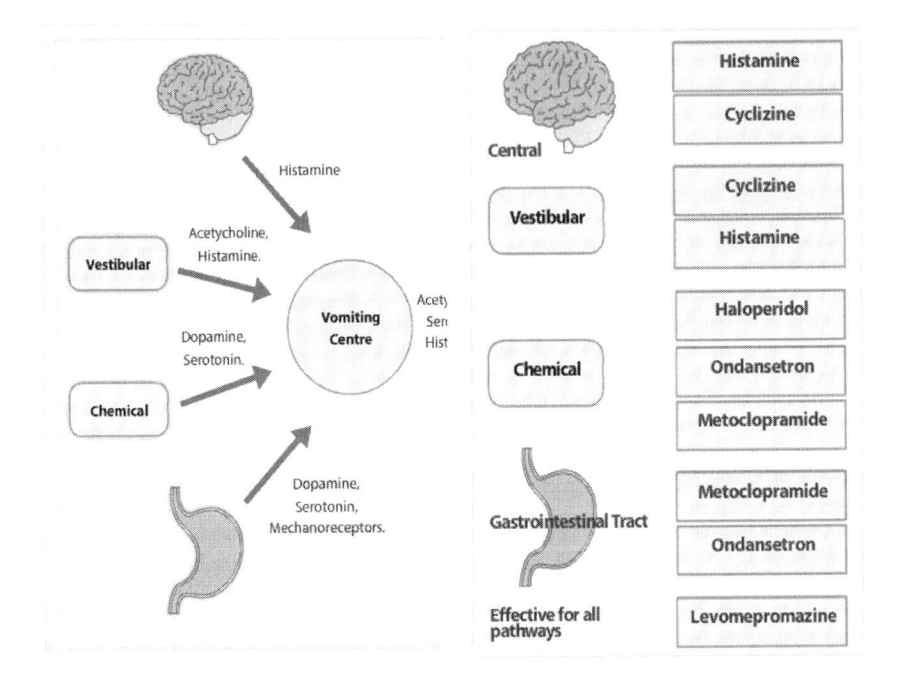

Practical assessment of nausea and vomiting in children

If nausea is not a problem:
- Think of oversensitive gag reflex, cough, post nasal drip, crying, gastric regurgitation), gastric stasis or compression, or possibly raised intracranial pressure

If vomiting seems to be associated with gagging or coughing:
- Think of post-nasal drip or reflux irritating the pharynx

If the child is otherwise well
- Think of gastric stasis, gastric distension, constipation

If there is nausea and the child is on drugs or generally unwell
- Think of drugs causes (cytotoxics, antibiotics, NSAID, opioids), infections or metabolic causes (e.g. renal or hepatic failure)

If the nausea is related to position/movement
- Think of vestibular infection/tumour, or vestibule-toxic drugs

If there is nausea and with features of oesophagitis or gastritis
- Think of drug related causes (NSAID or steroids) or gastro-oesophageal regurgitation or candidiasis

If there is nausea and anxiety
- Try to ascertain causes

If the child has a headache or neurological clinical features
- Think of raised ICP and/or intracranial infection

A practical approach to the of management of nausea and vomiting in children

All children
- Explain, reassure, modify diet to ensure it is frequent, low volume, and attractive.
- Address anxiety issues

Cough or gag?
- Treat any sinus infection or allergy, treat any reflux or chest infections

Possible iatrogenic causes?
- Review drug list and stop non-essential emetogenic drugs

Features of gastritis or oesophagitis (pain, indigestion, flatulence, blood in vomit)?
- Stop NSAID and steroids if possible, or minimize dosing. Use PPI or H2 antagonist with gaviscon (if tolerated).

Features of gastric stasis or dysmotility of the bowel?
- Try gastric motility stimulant (domperidone)

Related to movement?
- Rule out iatrogenic cause
- Start vestibulo-active drug such as cyclizine or hysocine hydrobromide

Related to toxic or systemic causes (e.g. infections, advanced tumours, renal or hepatic failure)?
- Treat underlying cause where possible
- Use drugs active on nervous system receptors, such as haloperidol, cyclizine

Due to cytotoxics?
- Consider ondansetron

Features of raised intracranial pressure?
- Consider radiotherapy or shunt. If not appropriate cyclizine is drug of choice. Consider dexamethasone.

Is the child anxious or agitated?
- Consider lorazepam, haloperidol or levomepromazine.

Above suggestions not working?
- Try low dose levomepromazine but do not sedate an actively vomiting child as you risk causing choking.

How do I manage constipation in children?

What you probably know already

- Constipation is common and distressing
- It can be treated using diet and drugs
- Drugs regimes can be planned rationally based on their different modes of action
    o Bulk forming laxatives like bran, isphagula, psyllum, methylcellulose.
    o Osmotic laxatives like magnesium salts, lactulose, sugar alcohols (sorbitol, mannitol, and lactitol), small volume enemas (citrates, phosphates), and polyethylene glycol.
    o Surfactants which act like detergents assisting water penetration into the stool, like docusates and poloxamer
    o Stimulant laxatives like bisacodyl, senna, sodium picosulphate.
    o Faecal softeners and lubricants like glycerine suppository, arachis oil, olive oil rectally.

What you may not know

- Rectal examination is rarely necessary and often distressing in children (rectal or anal tumours are very rare) and a loaded sigmoid can usually be picked up on abdominal examination.
- In the CPC setting constipation is commonly due to inactivity, dehydration, low food intake, fear of opening bowels due to pain, carers not addressing toilet needs regularly, neurological dysfunction of the GIT, metabolic derangements, and commonly used drugs (especially opioids, anticholinergics, iron and some anti-emetics).
- Constipation can be a trigger for seizures and dystonia in children with severe neurodisability.

Practical Assessment of Constipation in CPC

- Is the child dehydrated?
  - Think of reduced intake, or increased output (sweating, vomiting etc.).
- What is the diet like?
  - Is it simply reduced intake, or is the child eating 'easy-to-eat' carbohydrate based foods that are not high in fibre or stimulants?
- Is the child immobile?
  - If possible, encourage the child to be as active as possible.
- Is there nausea and vomiting with colicky pains?
  - Think of possible obstruction.
- Is there anal pain on defecation[52]?
  - Anal fissures are common in children and frequently lead to faecal withholding resulting in ever-larger and ever-harder stools, exacerbating the problem and setting off a vicious cycle.
- Are constipating drugs being used?
  - Especially opioids (the commonest cause in adult palliative care[53]), tricyclic antidepressants (e.g. amitriptyline), anticholinergic drugs (e.g. older antihistamines).
- Are there neurological features?
  - Neurological conditions often cause constipation[54] (e.g. HIV encephalopathy, spinal cord compression).
- Are there features of metabolic conditions?
  - E.g. renal failure, hypercalcaemia.

A practical approach to the management of constipation

Is there an obvious cause (see above)?
- Treat cause: diet, rehydrate, improve mobility, alter medications etc.

Signs of obstruction (e.g. vomiting, severe colicky abdominal pain, no flatus)
- Refer for immediate surgical opinion or, at end of life, stop feed, insert NG tube, consider IV fluid and use antispasmodic.
- Consider Octreotide if large volume gastric secretions

Pain on defecation?
- Check anus for fissures or candidiasis and treat with laxatives and local anaesthetic or glyceryl trinitrate cream prior to defaecation.

Hard stool?
- Start with movicol if available, or lactulose if not, increasing over one week. For movicol start at 2 sachets/d for 2-4y old and 4 sachets/day for older children and increase by 1 sachet per day up to a max of 8 in young children and 12 in older children.
- If no better add senna.
- If no better start rectal treatments (glycerine for hard stool, bisacodyl for soft stool).
- If still no better use micralax or phosphate enema.
- Treat with both stimulant and an either an osmotic or a softening laxative. Oral preparations are best, but rectal ground nut oil retained overnight may be helpful.
- If still no better manual evacuation may be necessary.

Manual Evacuation[55]

Manual removal of hard faeces may be necessary, particularly in resource poor settings with neurological causes of constipation. It is rarely now done in resource-rich settings, and in such a case the child would receive an anaesthetic. However, this may not be possible where you are. If not, the procedure is:
1. Explain to the child and family what is going to happen
2. Use distraction techniques.
3. Consider giving a mild sedative/muscle relaxant such as midazolam or diazepam.
4. Swaddle the child if young and it is acceptable
5. Prepare newspaper or other receptacle for the faeces

6. Apply KY jelly to gloved little finger
7. Stroke outside of anus to relax the sphincter, then gently insert finger, stopping if spasm occurs *giving time for muscles to relax*. NB Muscles do not relax instantly. Allow 5-10 minutes of gentle, firm pressure. Do NOT push hard against a closed sphincter. Be patient.
8. Remove small pieces of faeces piece by piece. Break up large pieces with finger before removal.
9. Talk to child throughout procedure and stop if discomfort is too much. You can always continue another day. This procedure may need to be repeated alternate mornings by a nurse or by a relative who can be shown the procedure
10. Once the "plug" has been removed a laxative at night can be commenced

How do I manage diarrhoea in children?

What you probably know already:

- Diarrhoea is very common in children, especially those with HIV/AIDS.
- It is caused by infection, drugs (especially antibiotics), constipation with overflow, bowel dysmotility, malabsorption, food intolerance (inherited or post-inflammatory) and ano-rectal irritation.
- In infectious diarrhoea, two mechanisms are at play:
  o increased intestinal secretion of fluid and electrolytes, predominantly in the small intestine.
  o decreased absorption of fluid, electrolytes, and sometimes nutrients that can involve the small and large intestine.

What you may not know:

- In resource poor settings it is the most common cause of illness and death during the first year of life with up to 70% of diarrhoeal deaths in HIV-infected children result from persistent diarrhoea[56].
- In HIV it may be due to other infectious causes (e.g. cryptosporidiosis, isosporiasis, CMV, atypical mycobacteria, strongyloides stercoralis, tricuris tricuria) as well as HIV enteropathy and side effects of antiretrovirals (especially protease inhibitors).
- The mainstay of all treatment is as follows
  o Supportive therapy—fluid and electrolyte replacement.
  o Treatment to reduce stool frequency and pain.
  o Antisecretory drug therapy to reduce faecal losses.
  o Specific therapy such as antimicrobial chemotherapy where needed.

Practical assessment of diarrhoea in CPC

- Are there signs of acute infection? (offensive or blood stained stool, pus in stool, fevers, abdominal pain)
    - Think retrovirus or other common infectious causes outlined above
- Is the diarrhoea chronic[57] in an HIV positive child?
    - Think rarer causes of infections (see above), lactose intolerance, medications (e.g., ARTs) & HIV enteropathy
- Are there signs of faecal impaction (previous constipation and palpable faeces per abdomen)?
    - Think overflow
- Could it be due to drugs?
    - Think antibiotics, ART's, excess laxatives
- Are there neurological features?
    - Think about autonomic dysfunction or other causes of dysmotility
- Is it related to diet and are there sugars or acid pH in stool?
    - Think about lactose intolerance or milk protein hypersensitivity
- Are there signs of ano-rectal irritation:
    - Think faecal incontinence, candidiasis, post radiotherapy

A practical approach to the management of diarrhoea

All children
- Use nappies or pads if available, or waterproof plastic under sheet covered with cotton sheet or absorbent material
- Counsel carers regarding cross-infection risks if necessary
- Counsel family regarding nutrition
- Counsel family regarding household hygiene (especially handling baby's water and food)

Diarrhoea – all children[58]
- Assess for malnutrition (see section on Feeding)
- Begin fluids immediately.

- Treat dehydration with oral rehydration salts (or IV fluids if severe)
- Emphasise continued feeding and increased feeding during and after the diarrhoeal episode.
- Abstain from administering anti-diarrhoeal drugs in acute diarrhoea.

Faecal impaction?
- See the section on constipation

Possible drug causes?
- Stop causative drugs if possible
- To reduce stool frequency and cramps
- Use anti-motility agent like loperamide or a diphenoxylate-atropine combination.

Diarrhoea with blood in the stool or child becoming systemically unwell?
- If stool tests are not available, treat presumptively according to your local guidelines or (if not available) as follows:
- Ciprofloxacin is the broadest spectrum antibiotic for invasive bacterial infection
- Cotrimoxazole/nalidixic acid and metronidazole should be used where amoebiasis is a possibility
- Metronidazole is useful where clostridium difficile is a possibility
- ciprofloxacin and metronidazole for children ≥ 8 years

Colic?
- Try antispasmodic e.g. hyoscine butylbromide

Proctitis or anal irritation?
- Local toilet
- Barrier cream
- Treat for candidiasis
- Metronidazole rectally if there is an offensive discharge
- Consider steroid suppositories or retention enemata

Justin Amery

How do I manage cough in children?

What you may know already

- Cough is common and irritating.
- It can be caused by infection, post-nasal drip, bronchial inflammation, oesophageal reflux, inhaled irritants, medications (ACE inhibitors) or airway tumours.
- It can be treated with simple measures such as steam, linctus and cough suppressants.

What you may not know

- Cough is a common symptom in AIDS, reported by 19% to 34 % of patients in surveys of symptom prevalence in HIV disease[59][60].
- Lymphoid interstitial pneumonitis and TB are common causes of cough in resource poor settings.

Practical assessment of persistent cough[61]
- Examine the child
    - Looking specifically for cyanosis (get peripheral oxygen saturation if possible), fever, respiratory rate, heart rate, stridor, wheezing, and in-drawing of chest/trachea
    - Chest X-ray if possible
- Barking, stridorous cough
    - Think laryngeal infection, foreign body or laryngeal pathology (larygomalacia, laryngeal stenosis, candidal laryngitis)
- Constant throat clearing, congested sinuses
    - Think post nasal drip, chronic sinus infection or inflammation
- Wheezing, no fever
    - Think of atopy/asthma or anaphylaxis?
- Productive cough
    - Think chronic infection, cystic fibrosis

- Very persistent, spasmodic whooping cough
    - Think pertussis
- Dry, breathless cough
    - Think interstitial lung disease
- Progressive cough with systemic symptoms such as weight loss, fevers
    - Think TB or progressive malignancy

A practical approach to the management of cough

All children
- Explanation
- Sit up, positioning
- Manage breathlessness if present
- Humidify air as much as possible
- Offer simple linctus
- Remember physiotherapy to assist drainage of secretions
- Stop smoking in the room, and also reduce use of stoves, kerosene lamps etc. in the house
- Consider cough suppressant such as codeine linctus or low dose morphine.
- Consider nebulised salbutamol. Even where there is no clear bronchospasm this can sometimes be relieving.
- There also also case reports of neublised lignocaine helping cough.

Infections?
- Treat underlying cause.

Bronchospasm?
- Nebulised salbutamol
- Nebulised ipratropium
- Oral or nebulised steroids

Lymphoid interstitial pneumonitis?
- Physiotherapy, pulsed steroids, bronchodilators, oxygen and ARTs

Recurrent aspiration and/or reflux
- See section on management of dysphagia and reflux

Post nasal drip?
- Position child upright
- Consider antihistamines
- Consider nasal steroid drops or spray

Irritating tumours in the bronchial tree?
- Use opioids and benzodiazepines to relieve symptoms of breathlessness and anxiety
- Consider nebulised local anaesthetic (e.g. lignocaine) if airway tumours but take care, these can inhibit gag reflex so do not feed for 1h before and after to reduce risk of aspiration..
- Consider radiotherapy.
- Consider heliox (which is easier to breathe against an obstructed airway)

How do I manage haemoptysis in children?

What you probably know already

- Haemoptysis is a very frightening event for the patient, the family and the professional.
- Nevertheless, panicking rarely helps anyone, so try to keep calm
- True haemoptysis (as opposed to bleeding from the upper respiratory tract) can be caused by infections, solid malignancy in the lungs or haematological malignancy, TB, aspergillus, pulmonary embolism, pulmonary hypertension, lung abscess and clotting disorders

What you may not know

- It is actually quite unusual for someone to bleed to death through the lungs.
- Most causes of apparent haemoptysis are due to non-serious bleeding from the nose, pharynx or upper oesophagus.
- Even true haemoptysis is not usually life-threatening, and will often settle once the cause of the bleeding is managed and the cough is suppressed (coughing generates tearing forces in the airways).
- That said, catastrophic haemoptysis does occur infrequently.

Practical assessment of haemoptysis

- Check it is real haemoptysis
    - In concurrent lung disease bright red frothy blood is likely to be from the lung, whereas dark or black blood, with coffee grounds, mixed with food in a vomiting child is likely to be haematemesis.
- Abrupt onset with cough, fever and purulent sputum?
    - Think acute lung infection

- Chronic, productive cough?
    - Think chronic lung infection
- Fevers, night sweats and weight loss
    - Think TB or malignancy
- Breathless, fatigue, frothy pink sputum, orthopnoea
    - Think heart failure
- Breathlessness and pleuritic chest pain
    - Think pleurisy or PE.
- Look for clues on examination
    - Especially fever, tachycardia, tachypnoea, weight loss, hypoxia, lymphadenopathy.
    - Inspect the naso-pharynx and oropharynx carefully
    - Look for other signs heart failure especially diastolic murmur, basal crepitations, tachycardia
    - Look for relevant signs of lung disease of infection such as unilateral or apical wheezes/crackles, signs of consolidation

A practical approach to the management of haemoptysis

In all patients
- Take a deep breath and try to stay calm
- Sit the child up and give oxygen
- If the child wants to lie down have the child lie on the side with the abnormal lung down.

Make a tricky ethical decision quickly
- Is this an end of life event requiring rapid symptom management, or is it in the child's best interest to treat aggressively?
- The prognosis of a severe true haemoptysis in the CPC setting is very poor, but, on the other hand, bleeding to death is frightening and traumatic for both the child and family, so if it can be stopped it may be worth aggressive treatment even if a child has advanced disease. See
- How do I deal with ethical decisions in CPC?
- How do I manage acute, distressing terminal events in CPC?

If you decide to treat aggressively
- Follow the 'ABC' principles
- Get IV access and give fluids
- Consider partial sedation with midazolam and opioid
- Consider blood and/or platelet transfusion if appropriate
- Consider haemostatic such as tranexamic acid
- If bleeding from varices consider octreotide
- Consider radiology and bronchoscopy if available
- Consider radiotherapy to tumour site if appropriate

Palliative care for catastrophic haemoptysis
- Give immediate terminal sedation: see section on 'how do I manage acute distressing terminal events'

If the child survives the initial bleed
- Treat the cause if possible
- Suppress the cough using opioids
- Consider sedation using benzodiazepines and opioids

Justin Amery

How do I manage breathlessness in children?

What you probably know already

- Breathlessness is a very frightening symptom, for the professionals as well as for the patient and family
- It needs rapid and assertive management
- It has many causes, for example mucus plugging, asthma, aspiration, recurrent chest infections, ventilator muscular weakness due to neuromuscular diseases, upper airway obstruction, diaphragmatic splinting and cor pulmonale, heart failure, and central nervous system dysregulation.
- Fortunately it responds quite well to non-pharmacological and pharmacological treatment

What you may not know

- Breathlessness is quite common in CPC[62][63].
- Dyspnoea, like pain, is a subjective sensation: there is not always correlation between what you can see (e.g. tachypnoea, cyanosis, respiratory effort) and the degree of dyspnoea felt by the child

Practical assessment of dyspnoea in children:

- Is the child blue?
    - o Think of obstruction (e.g. aspiration, upper airway obstruction, bronchospasm) or hypoventilation (e.g. debility, weakness)
- Fever?
    - o Think of infection
- Inspiratory stridor or a rasping sound?
    - o Think of acute upper airway obstruction
- Expiratory wheezing, tracheal tug, intercostal muscle recession?
    - o Think of bronchospasm/asthma

- Facial oedema or neck vein distension?
    - Think of superior vena cava (SVC) obstruction
- Orthopnoea (breathlessness on lying flat) and basal lung crackles?
    - Think of left ventricular failure
- Hyper-resonance and reduced air entry in the upper zones?
    - Think of pneumothorax
- Dullness and reduced air entry in the base?
    - Think of a pleural effusion

A practical approach to the management of breathlessness

All patients: non-pharmacological management
- Sit the child up if possible- they may be more able to breathe if they lean forward and use their forearms to support themselves
- Ensure some air flow into the room or across the child's face (using an open window, or fan)
- Give oxygen if signs of cyanosis or low pulse oximetry (if available)
- Reassure the child as much as possible
- Try and help the child relax using the techniques described in the chapter on pain
- Try and use the distraction techniques described in the chapter on pain

All patients: pharmacological management
- Give morphine and/or low dose benzodiazepines if the breathlessness is severe.
- Midazolam parenterally or buccally works fastest. Diazepam is slow to absorb, and no faster parenterally than rectally.
- If already on morphine, give a prn dose of 10-15% of the total 24-hour dose

Acute stridor or upper airway obstruction
- Take a deep breath and keep calm
- Carefully look in mouth to see if there is any easily removable foreign body, but take care not to push foreign bodies in deeper. If in doubt, leave and move on to next step.

- If you think there is a foreign body use Heimlich manoeuvre in older children, back blows and chest thumps for children less than 2yrs.
- Administer oxygen or heliox if available
- Consider nebulization with adrenalin (0.5ml/kg of 1:1000 solution, max 5ml) for 10 minutes.
- Consider tracheostomy if overall prognosis allows
- Consider dexamethasone high dose IV over 2 minutes
- Give parenteral morphine and/or benzodiazepine if the breathlessness is severe. (Oral/rectal diazepam works as fast as parenteral – you can use injectable diazepam rectally).
- Consider referral for surgery or radiotherapy or chemotherapy if available

Pneumothorax
- Oxygen
- Symptomatic treatment
- Consider chest drain

Acute heart failure
- Oxygen
- Loop diuretic (e.g. furosemide)
- Start or increase morphine

Chronic heart failure and cor pulmonale
- Oxygen
- Diuretics to reduce pulmonary oedema and breathlessness
- ART reduces episodes of intercurrent lower respiratory tract infections
- Consider inotropics, ACE inhibitors, digoxin

Bronchospasm (e.g. asthma)
- Oxygen high dose
- Bronchodilator nebulised or at least 10 puffs of metered-dose inhaler through a spacer device
- Parenteral or oral steroids

Pleural effusion
- Pleural drainage is effective in the short term, but fluid may recollect without pleurodesis, therefore you need to consider benefit v burden, especially with malignant effusions.

HIV pneumonia
- Unknown organism: chloramphenicol; or amoxicillin, flucloxacillin & gentamicin; or cefalosporins
- PCP: high dose septrin, oxygen, steroids

Lymphoid interstitial pneumonitis
- Physiotherapy, pulsed steroids, bronchodilators, oxygen and ART.

Justin Amery

How do I manage seizures in children?

What you probably know already

- Seizures are very frightening: for the child, for the family and for you.
- Most seizures are harmless and self-limiting and do not cause an imminent danger.
- If the child is safe, you don't necessarily need to rush straight to drug treatment. Just remove any potential dangers, lie the child on his/her side if possible, and give oxygen if available.
- Seizures are common in CPC (e.g. from brain tumours, neurological diseases, hypoxia, fever, raised intracranial pressure etc.).

What you may not know

- Time seems to run more slowly when you are witnessing a seizure so look at your watch as soon as it begins.
- Take a deep breath, try not to panic, and concentrate on calming the family and your colleagues while the seizure takes its course.
- If there is no improvement after five minutes, or if the child is turning blue, then you need to start treatment.
- Don't fall into the trap of thinking that the only successful outcome is to abolish seizures completely. This may be impossible.
- In a Children's Palliative Care setting it may be that a situation where a child is not necessarily seizure-free, but with seizures having minimum distressing impact on the child, might be a 'good-enough' outcome.

A practical approach to the management of seizures

All children
- DON'T PANIC: take a long, deep breath; breath it out slowly and check your watch

- Ensure the child is not in immediate danger (e.g. from falls, burns, drowning)
- Do not place anything in the mouth (e.g. spoons, tongue depressors)
- Give oxygen if available
- Check blood glucose level if possible
- Avoid leaping in with drug treatment but concentrate on calming family and colleagues
- Place in side-lying position after the seizure fit to prevent aspiration

After 5 minutes
- Give buccal midazolam or lorazepam (you can give the injectable form buccally if you don't have the buccal form available) or you can give it parenterally
- Alternatively give rectal diazepam but this is slower to work

No response after 5-10 minutes
- Repeat the dose

Still no response after another 5-10 minutes
- Give paraldehyde rectally - in an equal volume of oil

Still no response
- In most cases the family will want hospitalisation for IV infusion (usually of lorazepam, phenytoin or phenobarbital)
- In some cases, where the child still has a reasonable quality of life, anaesthetic and intubation may be required
- Ongoing management of seizures at this stage will be difficult to manage at home so usually the child will need admission.
- However, in rare cases, and if the family are keen for the child to be looked after at home then liaise with your local hospice or paediatric department to set up a subcutaneous infusion of phenobarbitone or phenytoin

Managing seizures at the end of life

- As with many other terminal symptoms, seizures at the end of life are often more distressing to the child's family than to the child his/herself. The child's family or caregivers need to be warned about the possibility of seizures at the end of life particularly in high risk cases and will need training in how to administer the drugs. Medication needs to be available in the home 'just in case'.
- Midazolam is the treatment of choice (as above). Otherwise, rectal diazepam or paraldehyde usually help to gain control over these seizures and may be administered regularly by the family themselves if the child is being cared for at home.
- Occasionally subcutaneous infusions (sci) of midazolam may need to be used. Although be aware that tolerance can develop and higher doses of midazolam can cause tachyphylaxis. In such cases phenobarbital or phenytoin should be prescribed.

How do I manage spasticity and dystonia in children?

What you probably know already

- Spasticity is tightness/stiffness of muscle whereas dystonia is sustained muscle contractions (spasms).
- They can be caused by genetic abnormalities, birth asphyxia, encephalopathy, infection, hypoxia, cerebral bleeding, drugs (e.g. metoclopramide, phenothiazines) or other causes of nerve damage.
- Muscle spasm is very painful (think of leg cramps or labour pains).
- Spasticity and dystonia can be so severe that they can lead to subluxation and dislocation of major joints (and imagine how painful that must be).

What you may not know

- Dystonia and spasticity often co-exist.
- Health professionals often overlook dystonia as a source of pain and other symptoms.
- Dystonia and spasticity often co-exist but have slightly different mechanisms which are important for understanding treatments.
- Spasticity can be caused by damage to the brain that reduces the activity of GABA (a relaxing neurotransmitter).
- Dystonia is due to damage to deep structures of the brain (such as basal ganglia, thalamus and cerebellum) which alters the activity of dopamine and acetyl choline.
- Apart from pain, spasticity and dystonia have other profound impacts on the life of a child including reduced or absent mobility (or even control over movements), profound fatigue, muscle contractures leading to bone and joint deformity and reduced/asymmetrical bone growth leading to small and/or deformed

stature, excessive secretions, drooling and feeding problems from a dysmotile GI tract.
- Dystonia can be triggered by anything that makes us feel 'jittery' (e.g. pain, cold, fever, anxiety, loud noises, full bladder or bowels, nausea).
- Because dystonia itself causes pain and anxiety, children can quickly get into a vicious cycle (pain or anxiety causing dystonia causing more pain and anxiety etc.)
- Spasticity and dystonia are tricky to treat and the drugs have many undesirable effects, which may limit their use.
- Non-pharmacological treatments are therefore even more essential than normal.
- Dystonia is not the same as epilepsy, but there is an overlap between the two and they frequently co-exist

Practical assessment of spasticity and dystonia

- Dystonia is a clinical diagnosis, so if you think a child may have it, and you are not sure, refer to a colleague.
- Aims of assessment
  o Confirm that dystonia is present and then to look for possible causes and co-morbidities, the functional impact, triggers and possible complications.
- Possible causes
  o This is usually obvious in the CPC setting but if not, go back through pregnancy, birth, early life, development, drugs, family history.
- Co-morbidities
  o Look for spasticity, joint problems, signs of gastrointestinal dysmotility, and seizures.
- Functional impact
  o Look at the section 'How do I assess all the needs of a child and family' for more details but an excellent question for assessing function is to ask the child or family to describe, in detail, their 'typical day' for a full 24 hours.

- Triggers
    - Anything that would make you feel tense or stressed is likely to be a possible trigger for dystonic spasm in CPC. Common triggers are pain, urinary retention, constipation, intercurrent illness, loud noises, bright lights, unfamiliar contact or positions, seizures
- Complications
    - Look for pain, gastro-oesophageal reflux, constipation, joint problems, dental decay, recurrent chest infections, anxiety and agitation.
- Examination
    - Try to characterise the dystonia, assess the level of functional impairment, and look for co-morbidities and complications, as well as looking for other movement disorders. Getting the family to video episodes is very useful as children may not show dystonia when you meet them.

A practical approach to the management of spasticity and dystonia

All patients
- Perform a full functional assessment and address problems with mobility, position, feeding, bathing etc.
- Listen to the family: every child is different and most families have worked out certain positions, tricks or tips that help break spasms
- Identify all community resources and coordinate a whole team approach
- Ensure gentle handling by familiar people in peaceful environments
- Teach the carer how to perform regular physiotherapy
- Warm bathing or hydrotherapy is often helpful

Treat causes
- Stop causative drugs
- Check that constipation or urinary retention are not an issue

- Ensure familiar, comfortable quiet environments
- Remove and treat all causes of pain as far as possible
- Try and control any epileptic activity

Analgesia
- Although simple analgesics are not very effective for dystonic pain, they are not necessarily ineffective and so are worth trying as part of an overall treatment regime, using the standard WHO approach.

Specific drugs
- In the author's experience, drugs for spasticity and dystonia show very variable and incomplete efficacy, and are rarely sufficient treatments in themselves.
- Drugs for dystonia may improve muscle tone but overall may impact the child's functioning or quality of lifemake things worse (e.g. spasm may be the only thing that allows a child to sit or stand)
- Drugs for dystonia may cause sedation and other adverse effects such as severe drooling
- Try baclofen and/or diazepam (or other benzodiazepine) first line as they may help with both spasticity and dystonia.
- Tizanidine and dantrolene can also help with both spasticity and dystonia as they are skeletal muscle relaxants.
- If spasms are no better, anticholinergics (such as benztropine) or dopamine blockers (such as tetrabenzene) can be useful
- Consulting with a neurologist is recommended if available

Physical treatments
- Depending on the site of dystonia, various local and physical treatments may be used
- Botulinim toxin injected directly into dystonic muscles can be very effective if available. It can take several weeks to reach peak effect.
- Intrathecal continuous baclofen infusion can be useful for generalised dystonia

- Surgical releases can help where spasticity or dystonia is causing painful or disabling deformity, subluxations or dislocations
- Braces, spinal fusion or rods do not treat spasm or spasticity but help reduce deformity of the spine (and consequent secondary effects). Be careful though as braces may actually create more pain if a child's spasms force his or her body against the brace.

Justin Amery

How do I manage skin problems in CPC?

What you probably know already

- Skin problems are very common
- Skin problems are very distressing, but may be overlooked or not taken seriously by health workers
- Most skin problems benefit from simple skin care advice
- There are simple and systematic ways of approaching management of skin problems; even if you do not know exactly what the underlying cause is.

What you may not know

- It is estimated that 92% of patients with HIV will experience skin problems during the course of the illness[64]
- The commonest causes of skin problems in HIV/AIDS in children are bacterial infections, fungal infections, viral infections (especially herpes) and infestations (especially scabies).
- As HIV/AIDS progresses, skin problems become more severe and problems such as candidiasis, oral hairy leukoplakia, folliculitis, herpes simplex and molluscum develop[65]
- Children with cancers can also suffer with fungating tumours, sores and hyperhidrosis (especially in lymphomas).

How do I work out what is wrong?

It is always better to look at pictures rather than read about rashes and spots. We cannot provide an atlas of rashes in this book, but you can find an excellent and free online resource (at the time of print) at *http://www.dermatlas.org/derm*

However, if you can't work out what you are managing, and have no one to turn to for advice, the following table will give you a reasonable chance of success.

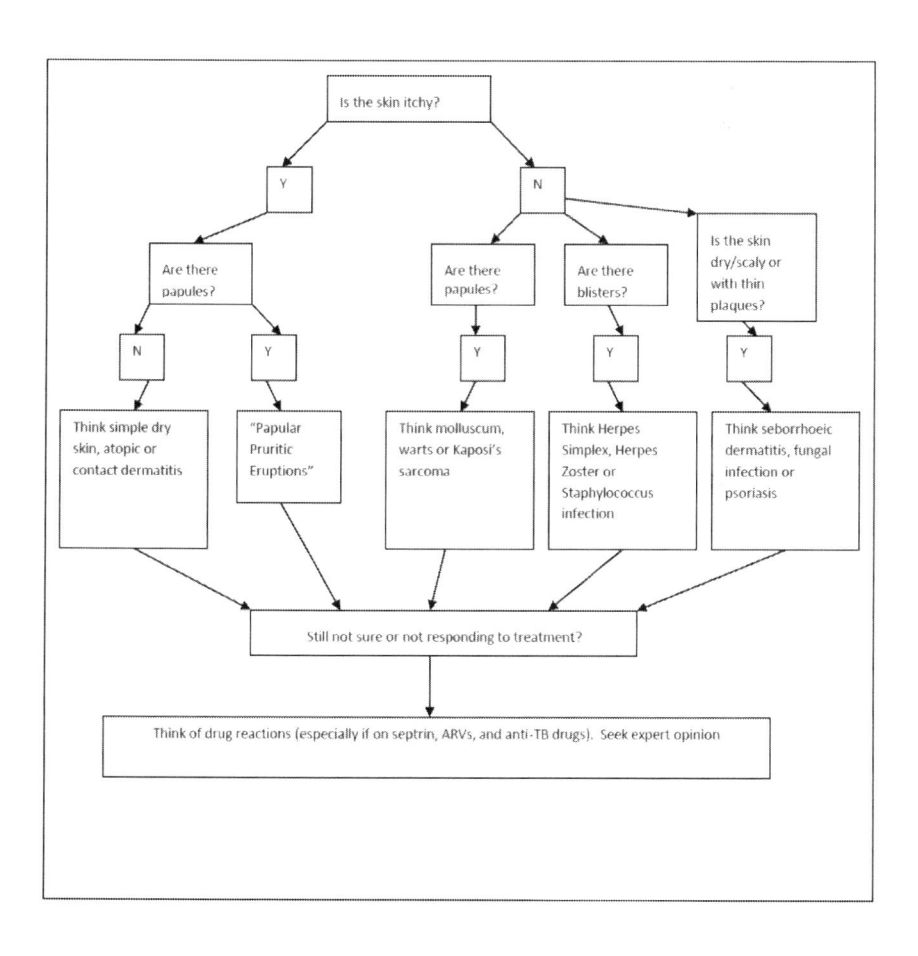

A Really Practical Handbook of Children's Palliative Care

Is the skin itchy?

- **Y** → Are there papules?
  - **N** → Think simple dry skin, atopic or contact dermatitis
  - **Y** → "Papular Pruritic Eruptions"
- **N**
  - Are there papules? **Y** → Think molluscum, warts or Kaposi's sarcoma
  - Are there blisters? **Y** → Think Herpes Simplex, Herpes Zoster or Staphylococcus infection
  - Is the skin dry/scaly or with thin plaques? **Y** → Think seborrhoeic dermatitis, fungal infection or psoriasis

Still not sure or not responding to treatment?

↓

Think of drug reactions (especially if on septrin, ARVs, and anti-TB drugs). Seek expert opinion

How do I provide general skin care?

Almost all skin conditions will improve to a certain extent with general skin care[66] so get into the habit of encouraging it with all families.

Intervention	Rationale
Avoid thorough washing, bathing or showering more than once a day. Don't use soap or baby bath, but substitute aqueous cream or emulsifying ointment or bath oil Use a thick moisturiser after bathing (e.g. aqueous cream, emulsifying ointment). Moisturise the whole body using thick moisturizer at least 3 times per day Humidify the air by boiling water.	Prevents drying of skin and removal of skin's natural moisture. Dry skin fissures and is therefore more vulnerable to infections and allergens.
Use soft, non-abrasive sponges, wash-cloths and towels. Pat rather than rub dry Keep fingernails short and smooth. Encourage the child not to scratch but to use finger tips to rub itchy areas Put mittens on small children to prevent damage from nails.	Prevents mechanical irritation.
Use lactic acid, urea (10 % urea cream), or sodium lactate moisturisers, or 20% salicylic acid for itching	Adds or helps to retain moisture
Keep topical creams and ointments cool or refrigerate.	The cooling sensation has an antipruritic effect.
Avoid restrictive or non-absorbent clothing.	Guards against mechanical irritation.
Use fragrance-free products rather than scented products.	Scented products may contain fragrance masking which elicit allergic responses in some patients.
Avoid lanolin-based creams.	Produces a high rate of allergic response.

How do I manage itchy skin?

Common Causes
The commonest causes in Children's Palliative Care are
- Simple dry skin (Xerosis)
- Atopic dermatitis
- Contact dermatitis
- Scabies
- Chicken pox
- Obstructive jaundice

All patients
- Basic general skin care (see above)
- Try to identify and remove allergens and irritants
- Use loads of emollients to trap moisture in the skin
- Use barrier creams (e.g. zinc oxide) where contact is inevitable (e.g. in nappy area)
- Urea containing creams may help
- Slightly sedating antihistamines, particularly at night to prevent itching, may also help (e.g. diphenhydramine or chlorpheniramine)

Itching
- Try wet wraps: damp cotton material soaked in potassium permanganate, water or saline
- Apply these over steroid ointment (for itch) and/or povidone-iodine (for infection)

Skin lesions
- Use adequate amounts of steroid cream (weak steroids to the face or flexures, moderate or potent steroids to thicker skin)
- Use "one step down" in infants and babies (start with hydrocortisone 0.5%)
- If infection suspected add flucloxacillin.

Seborrhoeic dermatitis
- Selenium sulphide shampoo

- Or antifungal creams
- Also topical steroids

How do I manage itchy papules (Papular Pruritic Eruptions – PPE)?

Itchy papules (papular pruritic eruptions) are very common in CPC, particularly in children with HIV/AIDS. Causes include:
- Insect bites and infestations (e.g. scabies, fleas, bed-bugs)
- Allergies (e.g. papular urticaria)
- Infections
- Folliculitis
- Fungal infections.

General treatment
- As for itchy dry skin above
- Topical steroids (start with hydrocortisone 1% and increase potency as needed)
- Topical antipruritics such as UEA with 1% menthol or calamine lotion or
- Consider sedating antihistamine such as promethazine or chlorpheniramine

Scabies
- Treat the whole family
- Wash all clothes and bedclothes
- Permethrin 5% cream
- Benzyl benzoate 25% apply & wash off after 24 hours – younger children dilute 1:2 in water)
- Gamma benzene hexachloride 1% lotion: apply & wash off after 24 hours
- Sulphur 5–10% ointment: apply daily for 3 days (better choice in young babies)
- Add flucloxacillin as staphylococcus co-infection is common

Folliculitis
- Treat empirically using agents active against fungal, bacterial and demodex mites as well as an anti-inflammatory agent
- E.g. clotrimazole or ketoconazole; erythromycin (topical or oral) and metronidazole 0.75% lotion or solution plus steroid (mid potency)
- Treat boils and furuncles with oral antibiotics (e.g. cloxacillin or erythromycin)
- If extensive or recalcitrant try oral antifungals

Fungal and Yeast Infections
- Treat with topical anti-fungals if localized,
- If extensive consider using systemic anti-fungals

How do I manage non-itchy scaly patches and plaques?

The commonest causes of scaly patches and plaques in CPC are
- Seborrhoeic dermatitis: superficial, non-itchy, greasy scaly rash, usually classically distributed on nasolabial folds, ears, eyebrows, forehead (but maybe much more severe and widespread in AIDS)
- Psoriasis: two types
  - Plaques: usually thicker plaques which have depth when you run your finger over them, usually on extensor surfaces.
    - Atypical: may present (especially in AIDS patients) with 'erythrodermic' (very widespread, painful, scaling inflamed skin) or 'pustular' (pustules developing in a widespread way, often associated with erythroderma)
- Fungal Infections: usually on sweaty areas, often asymmetrical, often starting small and gradually enlarging

All of these are much more common in HIV/AIDS

All patients
- General skin care as above

Seborrhoeic dermatitis
- Inflammatory lesions on the face can be treated with low-potency topical corticosteroid (e.g. hydrocortisone 1%)
- Antifungal agents such as selenium suphide shampoos, clotrimazole or ketoconazole topical cream can be very effective
- Dry, scaly skin can be managed with tar or salicylic acid-based shampoos
- Improves with anti-retroviral treatment in HIV infected children.

Psoriasis
- Topical mid-potency steroids (Classes I, II, or III) reduce inflammation
- Topical keratolytic agents such as tar and salicylic acid to reduce scaling
- Vitamin D derivatives (e.g. calcipotriol if available) reduce cell turnover
- Generalized disease may require systemic therapy which can cause problems with immunosuppression (e.g. methotrexate)

Fungal infections
- Keep body folds clean and dry and dust with cornflour or talc
- Initially try topical antifungal preparations for 2-3 weeks e.g. gentian violet 0.5% aqueous solution or Whitfield's ointment or imidazole creams, such as 1% clotrimazole
- Use oral antifungal agents if no response or widespread (e.g. griseofulvin, fluconazole, ketoconazole, itraconazole, or terbinafine)
- Beware potential drug interactions in patients on ARTs. You may need to monitor LFT's more frequently
- Tinea corporis or capitis usually requires 6 weeks or more of griseofulvin, nail infections take several months to resolve.

How do I manage non-itchy papules and nodules?

In CPC, most non-itchy papules are due to molluscum contagiosum or viral warts

Molluscum contagiosum
- Leave and reassure. If distressing local destructive methods can be tried, but often ineffective

Warts
- Topical 25% podophyllin or 50% trichloracetic acid, or local destructive methods if causing distress (but of questionable benefit).
- Condylomata accuminata (severe warty lesions in children with HIV/AIDS): usually improves with ART's and topical treatments as above, but may require destructive methods and sometimes even radiotherapy

How do I manage painful blistering lesions?

Painful blistering lesions are usually caused by herpes (simplex or zoster) in CPC. They are often preceded by pain, tingling and numbness in a dermatome, followed by an eruption of vesicles which burst open to form sores, eventually scabbing. In advanced HIV they may disseminate

Localised lesions
- Pain relief using WHO guidelines.
- Local antiseptics to try and prevent secondary infection. Anecdotally, the sap from cut frangipani leaves acts as a local anaesthetic
- Acyclovir topically for minimal lesions or orally in many lesions
- Give prophylaxis to immunocompromised contacts with varicella-zoster immune globulin (VZIG) 125U per 10 kg (max 625U) within 48–96 hours of exposure.

Herpes zoster
- As above. Normally systemic acyclovir is indicated.
- Give prophylaxis to immunocompromised contacts with varicella-zoster immune globulin (VZIG) 125U per 10 kg (max 625U) within 48–96 hours of exposure.

Widely disseminated lesions
- Consider hospitalizing in AIDS especially if disseminated, systemic or patient unable to take oral meds

How do I manage fungating tumours?

Skin tumours are relatively common in CPC settings, particularly in children with HIV/AIDS

All ulcerating lesions
- Cleanse regularly with saline
- Use topical antibiotic e.g. metronidazole gel (if available) or make up a cream (1 tub of aqueous cream, 40 x 400mg crushed metronidazole tablets and 10 mg of morphine powder highly concentrated)
- Decide if the ulcer is clean or necrotic and whether there is low, medium or high exudate, and choose correct dressing type from table below
- If these dressings are not available, locally made options includes ripe paw-paw for sloughing (crushed and applied twice daily for 5 days[67]), crushed charcoal (for necrotic and offensive wounds), live yoghurt, or honey.

Malignant tumours
- Consider radiotherapy and/or surgery

HIV related skin ulceration
- ART often results in tumour regression

The following list offers a rational approach to dressing usage in CPC

Clean, medium-to-high exudate ulcer (epithelialising)
- Paraffin gauze
- Knitted viscose primary dressing

Clean, dry, low exudate ulcer (epithelialising)
- Absorbent perforated plastic film-faced dressing
- Vapour-permeable adhesive film dressing

Clean, exudating ulcer (granulating)
- Hydrocolloids
- Foams
- Alginates

Slough-covered ulcer
- Hydrocolloids
- Hydrogels

Dry, necrotic
- Hydrocolloids
- Hydrogels

Table of dressing types (copied with permission from 'Dermnet' at http://dermnetnz.org/procedures/dressings.html

How do I manage hyperhidrosis (excessive sweating)?

Causes in Children's Palliative Care: Sweating is common in CPC and caused by HIV/AIDS, infections, lymphomas, TB and is also an opioid side effect

All patients
- Explanation and reassurance that there is nothing more sinister going on is often all that is required.
- Use frequent sponging; advise about absorbent clothing and bedding.

Topical treatment
- If available use aluminium chloride hexahydrate (ACH) to sweaty areas: noting it may take several weeks to reach maximum effectiveness. Use daily to start with, and beware it often causes local irritation which can put the child off using it, because it stings. If you see irritation beginning, stop the ACH briefly, use hydrocortisone twice daily until it settles, then restart the ACH again.

Drugs
- Drugs are not particularly effective but some patients find benefit with anti-muscarinics (e.g. glycopyrrolate, oxybutynin, amitriptyline). Where sweating is associated with anxiety, address the cause, in extreme cases propranolol or benzodiazepines (lorazepam) can be helpful.

How do I manage anxiety in children?

What you probably know already

- Anxiety is very common in children with life-limiting illnesses
- Children respond well to routine, openness, support and clear boundaries
- Children respond less well to change, uncertainty, blocked communication and stressful home and school life
- Simple things, such as adapting the environment to be more peaceful can make a big difference
- Talking therapies are as important as drug therapies

What you may not know

- Children with life-limiting illnesses are, in many ways, not different to other children.
- Their lives however have usually undergone profound change and uncertainty
- Even in the midst of a profoundly 'abnormal' situation, it is still possible, indeed vital, to try and normalise as much as possible
- Children are remarkably resilient, and can often cope with very difficult truths. They tend to be less good at dealing with deception or false reassurance.
- Your anxiety, and the family's anxiety, is catching for the child.

A practical approach to the management of anxiety in children

Listen
- It takes courage to sit down and listen, particularly if you are scared of what the child will ask or say.
- But the most helpful thing you can do is to listen deeply to the child, trying to become clear in your own mind about ideas,

concerns, thoughts, feelings and fears. This is deeply therapeutic in itself.

Normalise
- It seems strange to talk about normalization when a child is dying, but it is nevertheless crucial.
- It is natural t focus on the fact a child is dying, but in fact, despite that, there is much else in the child's remaining life that can be 'normal'. h So whatever you can normalize, normalize. Make life as predictable as possible, make clear your expectations, and don't change family routines and boundaries unnecessarily. Try and enable the child to stay at school (if he/she wants).
- It might sound strange but try also to 'normalise' the anxiety: anxiety is normal and adaptive. The child has a very serious illness and is possibly facing death. It is therefore both normal and OK to be anxious. It can be very helpful for a child simply to understand this.

Address unhelpful thoughts
- Because the child is dying, it seems odd to think that some thoughts are overly-catastrophic or unhelpful, but often they are. For example one child I looked after other was terrified of dying because she had a fear of worms. When we explained that she would not be awake or aware of the worms when she died, she became much less anxious.
- Discovering and challenging scary thoughts where possible or discussing and airing scary thoughts where necessary can help a child face his/her fears.

Teach useful behaviours
- There are a number of useful behaviours that you can teach or encourage. They will normally be things a child will choose to do, as they are enjoyable, like cuddling, playing, listening to or playing music, watching TV and so on. Try to make these happen.
- Older children and teenagers are also extremely responsive to breathing techniques, progressive muscular relaxation and guided

visualisation/relaxations, so look some up on the internet and give them a go.

Don't reassure falsely or excessively
- At a certain point, most children know that they will die, even if they are not sure what that means (who is?).
- False or excessive reassurance is therefore a bit like lying, and they know it. If they can't trust people their anxiety will get worse not better.
- This is not the same as suggesting you should jump clumsily in before they are ready, but once they start asking honest questions, gentle honest answers and suggestions for coping strategies are more effective than false reassurance.

Manage the whole family
- No child is going to be at peace in a home that is not at peace. This means we have to work with the whole family in a similar way: listening deeply, answering honestly, addressing unhelpful or false beliefs, suggesting helpful behaviours, and normalizing wherever possible.

Use formal psychotherapy
- Children respond well to art therapy, play therapy and music therapy – to such an extent that these should all be part of any CPC management approach, even if trained therapists are not available. See the 'Textbook of CPC in Africa' for more information on how to do this in resource poor settings.
- Older children respond well to cognitive behavioural approaches as well as more generic supportive counselling.

Use drugs
- In the early stages of palliative care, drugs are not usually that helpful, and may cause more side effects than benefits, although with older children anxiolytics (SSRIs), benzodiazepines or beta blockers may help.
- However, at the end of life, if a child becomes anxious or agitated, benzodiazepines or levomepromazine can both be helpful, though both are usually quite sedating.

What other resources might I find helpful?

- Goldman, A. Hain, R. Liben, S. Oxford Textbook of Palliative Care for Children, Second Edition. Oxford Textbooks In Palliative Medicine. OUP. 2012
- Together for Short Lives (2011) Basic Symptom Control in Paediatric Palliative Care – The Rainbows Children's Hospice Guidelines (9th Edition). [online] available from <http://www.act.org.uk/page.asp?section=167&sectionTitle=Basic+symptom+control+for+children%27s+palliative+care>
- WHO guidelines on the pharmacological treatment of persisting pain in children with medical illnesses http://whqlibdoc.who.int/publications/2012/9789241548120_Guidelines.pdf
- Amery, J. (Ed.) (2009) Children's Palliative Care in Africa. Oxford: Oxford University Press. Available for free download from: <http://www.icpcn.org.uk/core/core_picker/download.asp?id=204>
- Hain, R. and Jassal, S. (2009) *Paediatric Palliative Medicine. Oxford:* Oxford Specialist Handbook in Paediatrics
- Regnard, C. (Ed.) and Dean, M. (Ed.) (2010) *A Guide to Symptom Relief in Palliative Care Revised edition (6th Edition)* Oxford: Radcliffe Publishers
- Goldman, A. (ed.) (1994) *Care of the Dying Child.* Oxford: Oxford Medical Publications
- Children's BNF [online] available from < http://bnfc.org/bnfc/index.htm>
- Cancerpage.com (2007) *Pain Relief for Children [online]* available from <http://www.cancerpage.com/centers/pain/pediatrics_p.asp>
- Paediatric Pain Profile (n.d.) [online] available from <http://www.ppprofile.org.uk>

Part seven

How do I manage acute, distressing terminal symptoms at the end of life?

What you probably know already

- Emergencies can and do happen in any branch of medical care.
- The best way to manage emergencies is to avoid them – by good anticipation and planning.
- When they do occur, they are much easier to manage if you feel you are trained and prepared.
- Panicking rarely helps, can be paralysing and can be prevented by some basic rules.

What you might find useful

- When we use the term 'acute, distressing terminal events', we are referring to highly unpleasant symptoms, occurring at the end of life, and from which the child will die.
- The death of a child is highly distressing for the family; and no matter what you do, this will always be the case. In managing acute, distressing, terminal events, the objective is to minimise the suffering of the child so that the process of death is no more distressing than it has to be for the child or family.

Managing acute distressing terminal events: the golden rules

Mostly, where children's palliative care is available, children die without highly distressing physical symptoms. However, there are a number of distressing symptoms that may occur as a child dies, including severe pain, seizures, acute airway obstruction, massive bleeding and severe agitation. In such situations, the following five golden rules can be very useful:

1. Don't panic
If you find yourself with a child suffering from an acute distressing terminal event, you will probably feel very nervous, and perhaps out of your depth. However you feel, the child and family will feel much better if you look calm and in control (even if you don't feel that way inside).

Calm is catching, as is panic. Panic helps nobody, so try and stay calm. Momentarily drop your gaze. Breathe in, then all the way out. Imagine a quiet, protected place in your head or heart and put your feelings inside there. Drop your shoulders and unclench your jaw. Breathe in again, slowly all the way out, and then look up.

If you still feel like panicking, make an excuse to go outside for a bit (a mock phone call is always useful, or rummage around in your bag, or fetch something from the car) until you calm down.

Above all, remind yourself that:
- The child is dying and there is nothing you can do to stop that.
- It is a very sad experience for all, and you are not going to stop that either.
- It is extremely unlikely you will make things any worse.
- There are almost certainly symptoms you can control and comforting words you can say.

Even if you feel there is nothing you can do, and however useless you feel, just by being around you are helping and making the experience easier for the family, so stick at it.

2. Assess immaculately

Just as at any other time, the key to good symptom control is good assessment. At the end of life, things usually speed up and events can change quickly. That means you have to start assessing rapidly, even from minute to minute.

Your eyes and ears are important to assess what is happening around you, but so are your brain and your imagination. Think and imagine what is likely to happen next, how the child and family may be feeling, what you have got with you, what help might be useful, and what help is available.

Remember that sometimes, we are so desperate for a treatment to start working, we forget to leave it time to work. Seizures in particular seem to last forever. So when you have tried something, give it enough time before reassessing.

3. Hope for the best, prepare for the worst

Most acutely distressing terminal events are predictable. It's just that sometimes they catch you out when you are not prepared. The best thing you can do is go through all the worst case scenarios in your head in advance, practice exactly what you will do if that scenario arises, and make sure you, and everyone else, have everything they need to hand.

Try to arrange a meeting with the paediatric or hospice service involved (if available), to discuss and document the possible scenarios and agree a care management plan for each one. Most children who are cared for at home at the end of life have been discharged from hospital close to their death.

Make sure you (or the family) have a palliative care box available with drugs to be given immediately when such an event occurs. This should include parenteral midazolam (or lorazepam); parenteral diamorphine or morphine; a broad spectrum anti-emetic such as levomepromazine; and hyoscine patches.

If the child is at high risk of an acute terminal event (for example a child with a brain tumour may have terminal seizures, or a child with low platelets may bleed) this must be explained to the family and carers.

Although raising these events up front may be challenging, an informed family is usually an empowered family. It is also important to have these discussions with the family (and child if appropriate) to clear up any misconceptions about the reason for using drugs in this situation, as sometimes people can be under a misapprehension that drugs given at the end of life are being used as a form of euthanasia. While the child may and often does die rapidly after being given drugs for acute, terminal distress; this is non-intentional and probably coincidental, as by definition acute, distressing terminal events are terminal in themselves).

This is a fine ethical distinction, and one which is discussed more in the chapter on ethics, but it is a very important one in this context. If families are left feeling that they have authorised drugs to be used as euthanasia, it may lead to life long, but inappropriate, guilt. Therefore, in advance, check that the family all realise that if the drugs need to be used, it is only to ensure that the child's inevitable death is as peaceful and painless as possible.

4. Treat what you can treat
Although there are a number of acute, distressing terminal events that can happen, fortunately there is a generic approach which fits pretty much all these conditions. This is described below.

The main aims of treatment of distressing terminal events are to:
- Reduce pain.
- Reduce fear.
- Reduce the level of awareness of the patient where that is necessary to reduce distress.

5. Communicate
Fear of the unknown is always greater than fear of the known. The whole experience will be completely new to the child, and probably to the family too. They will watch you like hawks, constantly checking to make sure that you are able to guide them through the process with as little trauma as possible.

The child may or may not be conscious and aware at the time of their death. If he/she are aware, try to do everything you can to include and involve hi/her, so that he/she feel they have some choice and control.

In anticipation of a possible acute distressing event:
- Talk through in detail what might happen (best and worst case). For each scenario, explain to the child and agree exactly what the plan is:
    o For example, *"You might start to get breathless over the next few days. That is to be expected and we can help that by giving you some medicines. Do you want to ask anything about that?"*
- Try to involve the child in any plans so that they will keep active:
    o For example, *"If you start to feel breathless I need you to help me. Can you do that? Good. What I need you to do is to concentrate on those special breathing exercises I taught you. Shall we practice them again now?"*
- And again, for the family, talk through in detail what might happen (best and worst case). For each scenario, explain and agree exactly what the plan is.
    o For example, *"Andrew might start to get breathless over the next few days. That is to be expected and we can help that by giving him some medicines. Do you want to ask anything about that?"*
- Try to involve the family in any plans so that they will keep active:
    o For example, *"If Andrew starts to feel breathless I need you all to help me. Rachel (Mum), I need you to cuddle him so that he feels reassured, but not too tight. Try and make sure he can always see you. Tony (Dad), I want you to make a fan and gently fan his face so he feels that there is plenty of air around. Sarah (sister), I want you to help Andrew to think about something else by doing his breathing exercises with him, and maybe singing him some songs when he gets tired. Can you do that?"*

Then write a symptom management plan that the family and other practitioners can keep with them, so that everyone knows what to do (see 'How do I write a symptom management plan') for a template you could use.

A practical approach to treating any acute, distressing terminal event in CPC:

Sometimes we need to sedate a child rapidly, for example if there is acute upper airway obstruction or severe bleeding or seizures at the end of life. For all of these eventualities, the basic approach is the same: rapid sedation with midazolam (or equivalent) and morphine (or equivalent).

- Give buccal (or parenteral if buccal is inappropriate) midazolam 0.5mg/kg or the equivalent dose of diazepam rectally. Repeat every 10 minutes (for midazolam) or 20 minutes (for diazepam) until sedation is complete. For less acute presentations consider diazepam PR.
- Give buccal (or parenteral if buccal is inappropriate) morphine 0.2 mg/kg or the equivalent dose of diamorphine, which is more readily absorbed if available. Halve that for children under six months. Repeat every 10 minutes as required until sedation is complete.
- As soon as possible, set up a continuous subcutaneous or intravenous infusion of midazolam 0.3mg/kg/24hrs and morphine or diamorphine at a dose that is at least the equivalent of an intravenous breakthrough pain dose.
- Don't be scared to increase the doses rapidly as required. Remember there is no maximum dose of morphine for pain, and events can change rapidly during the end of life stage.
- Be prepared to anticipate and manage any side effects of morphine (nausea, vomiting, constipation etc.) that may occur.

It is important to explain to families that the sedative is not being used to hasten or cause the child's death but to relieve anxiety from what would be very distressing symptom.

It is also important to understand that the reason for rapid, terminal sedation is to prevent suffering. It is not a form of euthanasia, although by definition, children will usually die quickly as a result of the underlying disease process.

How do I manage restlessness and agitation at the end of life?

Terminal agitation and restlessness may be due to the underlying condition itself, or metabolic changes due to the underlying condition which you can unfortunately do little about.

Try and rule out the following:
- Uncontrolled hidden pain: especially bed sores.
- Urinary retention.
- Severe constipation.
- Hidden infection.
- Drug induced agitation

There is limited evidence around the management of delirium/agitation in palliative care. Consensus reviews suggest haloperidol is most effective for the hallucinations, and it also has a sedative effect, although not as powerful as benzodiazepines which can be used instead of, or alongside, haloperidol. An alternative, which has anti-hallucinatory as well as sedative effects, is levomepromazine.

How do I manage massive bleeding at the end of life?

Although rare, massive bleeding may be the terminal event in some conditions or their complications (e.g. Fanconi's Anaemia, bleeding varices in end stage liver disease, bone marrow failure etc.). Decisions should be made in advance as to when hospital admission and emergency haematological support are going to be withdrawn.

A massive bleed can be extremely frightening for the child and the family alike. Where massive haemorrhage is a possible mode of death, efforts should be made to get specialist support to avoid it (e.g. with platelet transfusions or tranexamic acid).

It is useful to give regular tranexamic acid to any child or young person orally for as long as possible if there is a possibility of terminal haemorrhage. In some countries, with support from your local paediatric unit, the child or young person may be able to continue on platelet support at home in the last few days. This would need to be given about twice per week.

In most cases of rapid bleeding the child loses consciousness quickly and does not suffer long. With less rapid bleeding consciousness may not be lost and the use of rapid terminal sedation (see above) may be indicated to relieve the child's anxiety.

It is common practice in palliative care settings to use dark sheets and pyjamas which makes the blood loss less apparent than if white linen was used. Nasal plugs or ribbon gauze should be available in palliative care settings as part of the emergency stock to manage epistaxis.

How do I manage acute severe upper airway obstruction (choking) at the end of life?

If the airway obstruction is due to an inter-current problem (e.g. inhaled foreign body in a child with swallowing difficulties) try and clear the airway using standard techniques for children (i.e. back blows for young children and Heimlich manoeuvre in older children).

Administer oxygen if available.

If the airway obstruction is irreversible and due to the underlying disease (e.g. tumour obstruction) the aim is to sedate the child as rapidly as possible.

If the breathlessness is severe and likely to be terminal, use rapid terminal sedation (see above) may be indicated to relieve the child's anxiety.

How do I manage noisy secretions ("death rattle") at the end of life?

All patients
- Position the child with his or her head low, so that secretions can drain from the mouth
- Ensure that any pulmonary oedema is excluded or treated with furosemide (be aware that a frequent cause is iatrogenic fluid overload near the end of life)
- If you suspect that the child this is more than just noisy secretions, but actually acute breathlessness, follow the guidance outlines in the section 'how do I manage breathlessness in children'
- If in doubt, treat for both.

Where secretions remain a problem
- Consider using hyoscine butyl bromide (Buscopan) or glycopyrronium subcutaneously to reduce the production of secretions
- Review after 30 minutes and repeat.
- You can repeat the doses every 4 hours or switch to a hysocine patch or subcutaneous infusion of hyoscine or glycopyrronium
- Only use suction in unconscious children and only very gently

What other resources might I find helpful?

- Goldman, A. Hain, R. Liben, S. Oxford Textbook of Palliative Care for Children, Second Edition. Oxford Textbooks In Palliative Medicine. OUP. 2012
- Amery, J. (Ed.) (2009) *Children's Palliative Care in Africa Chapter 18*. Oxford: Oxford University Press. Available from <http://www.icpcn.org.uk/core/core_picker/download.asp?id=204>

- Hain, R. and Jassal, S. (2009) *Paediatric Palliative Medicine. Oxford:* Oxford Specialist Handbook in Paediatrics
- Regnard, C. (ed) and Dean, M. (ed) (2010) *A guide to Symptom Relief in Palliative Care Revised edition (6th Edition)* Oxford: Radcliffe Publishers

Guidance

- Together for Short Lives (2011) Basic Symptom Control in Paediatric Palliative Care – The Rainbows Children's Hospice Guidelines (9th Edition). [online] available from <http://www.act.org.uk/page.asp?section=167&sectionTitle=Basic+symptom+control+for+children%27s+palliative+care>

Part eight

How do I play with children?

What you probably know already

- Play is the single most important way that children learn about, grow confident with and manage the stresses of living in their world
- Play is not just for fun. It provides developmental stimulation, distraction, exploration, socialisation and entertainment.
- In CPC, most children have suffered significant disruption and restriction to their normal family situations, so need play even more than healthy children
- Most health workers feel anxious about playing with children and lack confidence in how to go about it.

What you may not know

- Children are able to communicate unspoken anxieties and fears through the medium of play.
- Play is a therapeutic intervention for children who have experienced the shock of a life threatening diagnosis or traumatic medical procedures.
- Play therapy is a valuable addition to the holistic care provided to a child with a life limiting illness.
- Children with life-limiting illnesses are more vulnerable to developing learning disabilities and learning problems.

- Children with life-limiting illnesses are more likely than other children to miss out on play opportunities even though they need play to help them cope and come to terms with their illness.
- Even the sickest child can be helped to play.

The Key Principles of Play and Development

- Children are not simply 'mini-adults'. They see, feel and experience the world in a different way; use different thought processes to interpret it; and communicate and act differently to adults. If we don't understand this, we will find it very hard to understand children or be understood by them
- Children can develop at different rates down different developmental lines so it is important not to pre-judge a child's development
- Children learn from the way they see and experience their world; from the way people treat them; from what they see, hear and experience from the moment they are born.
- Children have 'critical learning periods' (Kandel[68]). We learn more in the first five years of our life than in all the rest of our lives put together!
- In the absence of relevant stimulation, parts of the child's brain become inactive, shrink down and eventually stop functioning altogether – permanently
- Many of the children you will encounter in day-to-day practice will not have been exposed to the necessary conditions in their environment to make the most of these critical learning periods.
- Brain stimulation children is not difficult: touch, hearing, sight – all the infant's senses can be stimulated by very simple methods such as talking, playing, holding and comforting the child, and meeting his physical and psychological needs.

- Play encourages physical, emotional, social, cognitive and language development. These skills are always important in childhood, but even more so when a child is ill and out of their normal routine.
- For children with life limiting illnesses, or for children who have experienced trauma and stress, play is the most natural means by which they can get rid of aggression, come to terms with the trauma of illness and impending death and attempt to take control of their world[69].

A brief overview of types of play

There are different types of play, and various ways of classifying them. For example[70][71]:

- Attunement play: The emotional attunement between infants and parents or others, for example through eye contact, smiles, talking/gurgling[72].
- Physical Play: The integration of muscles, nerves, and brain functions which help children to learning about their bodies and the world around them[73]
- Social Play: The interaction with others which help children learn social rules such as, give and take, reciprocity, cooperation, and sharing[74][75]
- Language and Narrative Play[76]: The taking part in language and stories which help children understand themselves, others around them, and their cultures.
- Constructive Play: The manipulation of the child's environment to create things, which allows a sense of value and accomplishment, the practice of problem solving and empowerment, and the development of dexterity[77].
- Fantasy Play[78][79] The acting out of new roles and situations, enabling children to experiment with language, emotions, ideas and challenges.
- Games With Rules: Which enable children to understand their part in family and societal contracts and rules..

A practical guide for playing with children in CPC

- Children from deprived settings, children who have been neglected or abused or children who have been sick for a long time may not be used to playing with toys, so you may need to show children initially what to do with some of the toys suggested below. However, take heart, most children need no more than a few seconds demonstration before they are trying to kick the ball, build a puzzle or colour the paper with as much enthusiasm as the others.
- Also, in many resource-poor CPC settings, money may not be available to buy expensive toys. But remember, children's rich imaginations are able to make pretty much anything into a toy and toys can be made out of many things we would normally discard We have tried to suggest toys below that can be easily made or are readily available as day-to-day household implements.
- Remember that toys need to be safe and clean. In most children's palliative care settings there is a cross infection risk, so toys should be kept clean and well maintained and thrown out if they break or become unsafe.

Playing with children from 0 - 18 months

- Suitable toys: Mobiles, pictures, music toys, rattles, stacking toys, pots and pans, plastic containers, large puzzles, materials of different textures to squeeze and chew, toys to pull on a string (e.g. painted plastic bottles)or to bang together (e.g. spoons and pots), large dolls
- Play ideas for care givers of 0 – 18 month old children: Respond to baby's sounds, make eye contact, concentrate on social and gross motor play, smile, let the baby play with your fingers, talk and sing, play "peek-a-boo," and hiding games, make faces, play at "losing" and "finding" things, name objects, shake a rattle, use objects with different textures to touch, encourage him to crawl to toys out of his reach.

Playing with children from 18 months to 3 years

- Suitable toys: Things to ride on, push-pull toys (such as wire cars), sandpit and sandpit toys (such as spoons, funnels, plastic bottles and tins with lids), weaving and beading materials, crayons, paints and chalk; wooden blocks, packing boxes for climbing in and out, dress-up clothes, soft toys and dolls, kitchen implements, picture books, music and story CDs, simple musical instruments and play dough (you can make your own)
- Play ideas for care givers of 1-2 year olds: encourage movement, hide things, catching and chasing, building blocks, follow-the-leader, guessing games, act out stories, copying household chores, reading and singing, story telling, singing, clapping and dancing, naming games, bubbles blowing, playing with balls, drawing and colouring.

Playing with children from 3 to 6 years

- Suitable toys: Climbing equipment such as ropes, swings, tyres, earth mounds; balls and bats; dolls and soft toys; large puzzles or board games; toy vehicles (bought or made with wire); rhythm instruments like drums; threading beads, blocks, buttons; dress-up outfits; construction sets; crayons, paints, pencils; play dough; gluing; sand play; story books and CDs; puppets.
- Play ideas for care givers of 3 – 6 year olds: out and about physical play; make-believe telephone conversations; hide-and-seek; singing and dancing; card games and board games; counting" and "number" games; puppets; reading; naming; play dough; dress up; 'pretend play' with household implements; sing songs with movements; sand and water play; crayons, paint and paper; balls and chasing games

Playing with children from 6 – 12 years

- Suitable Toys: scooters, skateboards, carts; ball sports and games; kites; board games; dolls and soft toys; toy vehicles;

- construction materials; puzzles; disused computers and keyboards; dress –up outfits; miniature people and vehicles; art materials; books and music and (of course – if available) computer and console games
- Play ideas for care givers of 6 – 9 year olds: Be observant; provide opportunities for make-believe games; encourage constructing and creativity; help organize and classify things; allow the children to play competitively at games and play situations; use puppets to tell stories; play language games e.g. ask children to find "rhyming" words or synonyms; encourage creative writing and poetry; singing and dancing games; team sports; jokes and riddles; reading (them to you and you to them)) and outings to interesting places

How do I set up a children's play area?

It may be that you have very little space in which to operate. However, in case you do have a suitable area, we have suggested below some ideas for how you might set up a play area for the children in your care.
In general terms it is good to have a
- A cognitive (thinking) area: for children to think and learn
- A book area: for children to read and be read to
- A fantasy area: for dressing up and puppet plays
- A creative area: for painting, drawing, creative play and crafts
- An outdoor area: for energetic play
- Sleeping area: for sleep and rest

What materials can you use for play in resource poor areas?

If you are in a resource-poor area, it may be difficult to access a lot of toys. That's not a problem. Children will find ways of playing in just about any situation. Older children are also experts at making toys – so consult the experts before doing anything else. But here are a few tips that might help you get started

- Powder paint can be mixed with wallpaper glue and water to make it go further.
- Crayons are just about the best investment you can make if you only have a very small budget
- Get old magazines for cutting out and pasting
- You can make glue by mixing flour and water to a paste
- Old boxes, cartons, lids, are great for all sorts of play
- Singing, dancing and clapping are free, while finding something for a child to bang is also usually easy
- Most children love to play with safe household implements
- Shakers can be made with plastic yoghurt cartons – put some maize rice, stones or plain rice inside and cover with a large sticker or paper and elastic band. Shake in time to the music.
- Make your own play dough
- Make a ball from disused plastic and paper bags and string
- Use old ropes for skipping
- Create a messy play area outside in a unused piece of land.
- Add some sand if you can access it

Resources that you might find helpful

- UN Convention on the Rights of the Child Article 31 www2.ohchr.org/english/law/crc.htm. The right to play and informal recreation is a human right for children and young people up to 18 years of age, enshrined in the UN Convention of the Rights of the Child.
- Best Play - What play provision should do for children www.playengland.org.uk/Page.asp?originx_837hq_60550341588120v64o_20081065015y This report looks into the benefits of play for children, as well as the consequences of inadequate play provision.
- Charter for Children's Play www.playengland.org.uk/charter. The Charter outlines eight statements or principles that describe a vision of play.

- Free play in early childhood: A literature review. www.playengland.org.uk/Page.asp?originx_9871bw_102416424017g62k_2008624223n. A comprehensive literature review focusing on free play and the early years.
- Key sources of information on children's play www.ncb.org.uk/cpis/resources/factsheets.aspx. Compiled by Children's Play Information Services (CPIS), this factsheet lists some of the key organisations and sources of information on children's play.
- Playwork Principles: www.skillsactive.com/playwork/principles. These principles establish the professional and ethical framework for play work. It describes what is unique about play and play work, providing the play work perspective for working with children and young people.
- Why Play? www.sustrans.org.uk/resources/publications. This leaflet looks at the benefits of active play and how to make it part of everyday life.

Part nine

How do I provide palliative care for adolescents?

What you probably know already

- Adolescence is time of rapid change and challenge: physical, psychological, spiritual and sexual
- Adolescents form a distinct group with their own unique physical, emotional, psychological and social needs that are significantly different from those of adults and children
- Adolescence is a time of wide mood swings, intense feelings, higher risk taking, lower impulse control, broad role exploration, increased sense of vulnerability, developing sexual awareness and the movement from concrete to abstract thinking. All of these affect how adolescents experience life-limiting illnesses

What you may not know

- The numbers of adolescents requiring CPC worldwide is huge
- Adolescents may not fully understand their disease, its prognosis and its effects on self and family.
- Adolescents may not have formed fully developed adult defence mechanisms to help cope with the illness and its implications[80] and so may use defence mechanisms such as anger and withdrawal that increase their sense of isolation and make care trickier to provide.

General principles of palliative care for adolescents

Palliative care should aim to help adolescents to[81]:
- Have the best possible quality of life despite their illness.
- Be in control as much as possible.
- Be clear about their personal identity.
- Accept their new body image.
- Gain and maintain some freedom from their parents without losing their support and open communication.
- Develop a personal value system.

These aims are difficult to achieve in health-care environments because of a number of factors, including[82]:
- The restriction of physical activities due to ill-health and health-care environments.
- The threat of change in body image posed by illness and treatment.
- The lack of privacy.
- The threat to life of illness and treatment.
- The use of anger, withdrawal and denial as defence mechanisms
- The change of environment and the effects of illness and treatment cause loss of independence.
- Professionals being caught between parents (who want to protect) and adolescents (who want independence and respect).
- Separation from peer groups which can generate fear of rejection and undue dependence upon family and care professionals.

It is therefore important that you:
- Involve them as much as possible in decision making
- Maximise privacy
- Encourage self-care
- Provide honest and realistic information about the effects of illness and treatment on body image, and support them through those changes
- Encourage and support peer groups

- Explore fears, anger and other emotions in order to help boost their sense of control and independence.
- Maintain schooling and other 'normal; activities as far as possible

The illness experience of adolescents

The illness experience of adolescents will vary according to important influences on their life which may include: [83] [84] [85]

- An increased need to preserve normality (e.g. school attendance etc.) even if this seems foolish to adults. Disrupted schooling and education is a particular source of anxiety.
- An increased gap between the perception of their actual body image and their ideal body image due to the physical effects of illness.
- The perception of the loss of personal control is exacerbated by illness and worsened by the tendency of parents to become over-protective at such traumatic times.
- Anxiety and uncertainty about the future are heightened in illness.
- Adolescents' psychological defence mechanisms often include denial, withdrawal and anger. These all mitigate against good and open communication, thus exacerbating the sense of loss of control and increase anxiety and uncertainty.
- Partly as a result of these factors, conflict may develop, often between adolescents and parents but sometimes between adolescents and other carers.
- Illnesses lead to changes in social relationships at a time when peer group support and acceptance are particularly important. The maintenance of support from and contact with peers is crucial[86].
- Questions such as the meaning of life and the reasons for pain, suffering, unfairness and punishment assume much greater significance when the adolescent has a life-limiting illness[87]. Therefore palliative care for adolescents needs to include spiritual care and allow the adolescent to explore such issues, including their lost hope for the future.

Communicating with adolescents in CPC

The challenges of communicating with adolescents in CPC
- It takes time to build trust and communication.
- If adolescents are not aware of their illness and included in decision making there is more chance of them defaulting or not adhering to treatment.
- They are less likely to seek health care from other sources so if they are not given the opportunity by health workers to explore the meaning of their illness then they are unlikely to get this opportunity elsewhere.
- Lack of cognitive development can result in a reduced awareness of the consequences of the illness and decision making.
- Adolescents have a need for acceptance from health workers even in the face of inappropriate behaviours.
- Adolescents have a need for balance between structure and freedom.

Communication can be improved by[88]:
- Creating an adolescent friendly environment.
- Establishing rapport through respect and acceptance.
- Taking time to build trust and relationships before dealing with difficult issues.
- Using activities, such as art and music therapy, to help them explore feelings and difficult issues.
- Being open and honest about illness, death and dying and not making promises that you may not be able to keep.
- Assessing risks and discussing these openly.
- Ensuring accurate and relevant information is available and provided to the adolescents when they are ready to hear it.

Factors which inhibit good CPC in adolescence:[89,90]
- Low expectations for themselves and for their care package.
- Lack of trust in professionals as a result of perceived "failure" of previous care.

- Lack of knowledge of existing educational and vocational opportunities.
- Lack of self-advocacy skills.
- Low self-esteem.
- Lower levels of integration with peers.
- Heightened orientation to and dependence upon adults.
- Low educational aspirations.
- Poor knowledge of sexuality.

There are particular challenges of adolescent CPC in resource poor settings, for example:

- Many adolescents head their households, often caring for siblings, parents or other relatives. These inhibit peer group support, education, and the development of independence.
- School dropout: Children who leave school are less likely to develop the skills necessary to abstain from sex or practice safe sex; they are economically vulnerable and open to sexual exploitation.[91]
- Financial problems: which may lead to family break-up; with consequent risks to health and safety of all children and adolescents.
- Access to health care: may be poor, so potentially treatable problems may go untreated; thereby increasing the burden of the disease.
- HIV/AIDS and cancers often stigmatise at the time when stigmatisation is most painful
- HIV/AIDS and cancer often lead to adverse body changes just at a time when body image is the least settled and the most worrying.
- Adolescents are particularly vulnerable to sexual abuse and exploitation[92,93,94,95]

Adolescence and adherence to treatment

Adolescents are less likely than adults to adhere to treatment, which is a problem in CPC, but especially in HIV/AIDS, when even occasional loss of adherence to ARTs may lead to drug resistance and therefore premature death. There are many reasons for this[96]

- Adolescence is a time of rapid development and uncertainty, so it can be difficult to follow strict routines.
- Drugs are a visible sign of being different at a time when stigmatisation and peer pressure are particularly powerful.
- Adolescents often have a low self-esteem and may well be grieving for the loss of loved ones. These can both lead to the feeling "what does it matter if I take them? I'm not really worth it anyway".
- Use of alcohol and other substances becomes common in this age-group. Alcohol and street drugs make users more chaotic and may adversely interact with medications.
- Fear of the disease and of the treatment may cause adolescents to go into denial; and stop taking the drugs.
- Lack of understanding: health education and access to health information may be reduced, leading to a lack of awareness of consequences of defaulting.

Strategies for maximising adherence are fundamentally aimed at giving as much control, information and autonomy to the adolescent as possible; thereby enabling him or her to make informed choices and also demonstrating that you trust him/her, which can boost self-esteem and confidence. Tips would include

- Making it clear that you are giving responsibility for medication to the adolescent
- Give plenty of age-appropriate education and information
- Get them to document their medication: perhaps asking them to design and develop their own diary for the purpose
- Set up support structures such as school teachers, parents, carers, friends
- Ensure disclosure to key supporters so that they can keep a distant and supportive eye out
- Connect to support groups if they are available
- Give plenty of positive feedback and encouragement

Care of dying adolescents

The aims of palliative care remain the same for dying adolescents as for that for children and adults: enabling them to die in peace, comfort and dignity. Similarly, the basic practice of providing palliative care to adolescents is also the same as in children and adults. In other words establish good communication and rapport, perform a thorough and holistic assessment, draw up a problem list of physical, psychological, social, spiritual and financial concerns, and draw up and agree a management plan to address each of these problems.

However, there are some specifics pertaining to caring for the dying adolescent[97]

- Clinical depression is prevalent amongst dying adolescents (17%).
- Most dying adolescents are aware of the prognosis, yet often this is not discussed.
- Adolescents are capable of discussing their prognosis, but need to be able to set the pace of such discussions.
- Hospitalisation and illness reinforce social isolation. This can lead to prolonged dependence of adolescents upon their parents.
- In order to counter the tendency of parental over-protection, the dying adolescent should be given opportunities to spend time and develop relationships outside the immediate family group.
- Adolescents have a great need to explore their emotions and fears, so a "permission giving" atmosphere and staff who can facilitate this are vital.
- Practical issues such as will-making, funeral planning, and writing good-bye messages seem to be very useful.
- Sibling relationships in this age group are often very strong, so good sibling care is important.
- There are frequently many professionals and carers involved in the care of a dying adolescent. Adolescents will however often select only one person with whom they will communicate, therefore adequate communication between carers can be a problem.

When setting about care-planning for adolescents with life-limiting illnesses; we recommend that you try and follow the following principles:
- Adolescents tend to need to seek and find reasons, and are happier when they have them. Therefore try and help adolescents to find reasons and meaning.
- Give as much of control over what is happening as possible (e.g. involving them in decision making during life and helping them to plan for their death).
- Allow adolescents to ventilate and explore their concerns and fears[98].
- Facilitate and encourage social contact[99] but give "permission" for adolescents to withdraw as death approaches and facilitate communication and sharing amongst their chosen intimate family and professional contacts at that time.
- Give plenty of opportunities for discussing prognoses. Most dying adolescents are aware of the prognosis, yet often this is not discussed.
- Sibling relationships in this age group are often very strong, so ensuring good sibling care is important.[100]
- Adolescents often use symbolic rather than direct ways to communicate their feelings and concerns: so "play" or occupational therapy is as important as in younger children. Drama, poetry, music and art are particularly appropriate for these age groups.
- Adolescents are able to think abstractly and symbolically so music, drama, poetry and art are particularly helpful
- Adolescents like to seek meaning; reasons and patterns which means that opportunities to learn and take part in discussions (especially in peer groups) are crucial.

Practical activities that you may find useful

- Making a memory box or book, to help them find ongoing meaning and connection.
- Making a family record to help them gain a sense of where they fit into the family; and therefore their 'legacy' within the family after they have gone.

- Draw up an adolescent's 'will': they may have few possessions, but it is important to discuss thoughts about what they want to happen after death and how they wish to be remembered. All of these things can be captured in a will.
- Involve and encourage peer groups to learn about and discuss factual topics about the disease process and its treatments; to consider broader issues such as sexuality and adherence; discuss and share experiences of illness and health in order to enable mutual support; and to ensure that adolescents also make time to have fun.
- Facilitate symbolic play, art, poetry, music and drama; all of which offer distraction, enjoyment, ways to learn about their condition and ways to express their feelings

What other resources might I find helpful?

- For teenagers with HIV/AIDS - http://www.pozitude.co.uk/
- For teenagers with cancer - http://www.teenslivingwithcancer.org/ and https://www.teenagecancertrust.org/who-we-are/about-us/
- 'Palliative Care for Young People Aged 13 – 24'. September 2001. Researched and written by Rosemary Thornes on behalf of the Joint Working Party on Palliative Care for Adolescents and Young Adults. Edited by Stella Elston http://www.palliativecarescotland.org.uk/content/publications/PalliativeCareforYoungPeopleaged13-24.pdf
- Goldman, A. Hain, R. Liben, S. Oxford Textbook of Palliative Care for Children, Second Edition. Oxford Textbooks In Palliative Medicine. OUP. 2012

Part ten

How do I deal with ethical dilemmas in children's palliative care?

What you probably know already

- All professionals have a duty to:
  o Respect the life and health of their patients.
  o Perform to acceptable standards.
  o Maximise benefits for patients and to minimise harm.
  o Respect a patient's autonomous wishes.
  o Act rationally, honestly, fairly and professionally.
- Competent patients have "...*an absolute right to choose whether to consent to medical treatment, to refuse it [which] exists notwithstanding that the reasons for making the choice are rational, or irrational, unknown or even non-existent". (*Re T 1992)[101].
- For incompetent patients, the proxy decision-making team (including the patient's next of kin) must reach consensus agreement regarding a management strategy that is in the child's best interests.
- Competence is situation specific. It depends upon the ability of the patient to assimilate, understand and retain the information that is provided by professionals. The determination of competence is complex and a subjective value judgement.
- The doctrine of necessity permits professionals to intervene without consent in a life-threatening emergency.
- Professionals do not have a duty to carry out medical treatment against their personal professional judgement.

What you might find useful:

Are ethics different with children? Yes and no. Ethical principles are the same, whether applied to adults or to children, but the *application* of the principles varies because children:
- May not fully understand their illnesses and treatments.
- May not be able to communicate fully their thoughts and wishes.
- Have the potential to become autonomous adults and acquire competence.
- Their rights may be less well respected as those of adults.

In applying ethical principles to the practice of children's palliative care, there are three key questions to answer:
1. What needs to be decided?
2. Who decides?
3. How do they decide?

What needs to be decided?

This is probably the most important step in trying to manage any ethical dilemma. If we are not clear about the question, it is very difficult to come up with any useful answers. For example, in the case of a child with end-stage oral cancer, there may be several different dilemmas, such as:
- Should we disclose to the child that the end stage of life has been reached and, if so, when?
- Should the child be cared for at home, a hospice or in hospital?
- Should the child continue to be fed, and if so, how?

If you are 'stuck' with a case, it often helps to list out the various different questions/dilemmas, and prioritise which need deciding when. It may be helpful to consider the intention of each element of treatment: comfort care (symptom control and palliation), invasive therapeutic care, or both, and then balance as to whether the potential of achieving a beneficial gain of each facet of care outweighs the risk of imposing a burden.

Who decides?

This varies from country to country. In my own country, England, children over the age of 16 are presumed to be 'competent' to give informed consent or dissent to proposed medical treatments[102].

According to the 'Fraser Guidelines[103]' in the UK, a child below the age of 16 may still be competent to give informed consent: *"...whether or not a child is capable of giving the necessary consent will depend on the child's maturity and understanding and the nature of the consent required. The child must be capable of making a reasonable assessment of the advantages and disadvantages of the treatment proposed, so the consent, if given, can be properly and fairly described as true consent"*.

Even if a child is not fully competent, this does not mean they should be excluded from the decision making process, as they may be at least partly competent.

The Royal College of Paediatrics and Child Health (UK)[104] has laid down guidelines suggesting that, whether the child is legally competent or not, wherever possible professionals should attempt to ensure that children are informed and consulted about the relevant decision and that their views are respected and taken into account in decision-making.

This seems like a just and reasonable approach, wherever you may live.

Who decides if the child is not competent?

The child's best interests are always paramount, and all factors including medical, emotional and other welfare issues must always be considered.

In most situations the parents, or those with parental responsibility, would have the responsibility for making decisions for their child. However, in healthcare decisions, in most countries the rights of parents yield to the child's best interest test. There are occasions where the parental wishes do not seem to be in the child's best interests, and in that situation the regulations vary from country to country.

It is ideal if the multi-disciplinary team works with the parents to try to reach a consensus agreement. No single opinion should be decisive; however in most countries the senior physician has the overarching responsibility to decide where a consensus cannot be reached. The decision should be based on a reasonable judgement as to what is in the child's best interests. Please check your own country's legislation on this. If there is significant dispute, and if time allows, it is often wiser to seek a legal opinion.

How do we decide?

Once the dilemma has been clarified and the question of who should be involved in the decision has been settled, the final step in the ethical decision making process is to decide how to decide.

In ethical theory, decisions can be based upon:
- Beliefs: e.g. the belief that all life is sacred.
- Duties: e.g. the duty to act in your patients best interests.
- Consequences: e.g. the premise that it is best to act in a way that will be likely to do more good than harm.
- Values: e.g. that health professionals should act out of compassion, honesty and fairness.
- Rights: e.g. that the child has a right to the best possible treatment.

What are the problems applying ethical principles in children's palliative care?

- There are a number of ethical principles that are relevant to a given clinical scenario.
- The decision-making process can be fraught with controversy.
- Individuals subscribe to different ethical theories with variable conviction, potentially producing a plurality of moral beliefs and assumptions.
- Stakeholders may attribute variable and inconsistent weighting to the numerous competing interests and influences.

- Personal experience, professional and social background, religious and cultural perspectives and emotional status all influence an individual's values and beliefs.
- Quality of life appraisals are highly subjective, speculative and the subject of recurrent academic criticism.
- Professionals may be under conflicting pressures which may, consciously or subconsciously, influence their analysis (e.g. cost, targets, time, and personal exhaustion).
- Making objective and rational choices in emotionally painful and taxing situations is not easy, as empathy and psychological defence mechanisms invariably come into play.

Deciding what to do in practice

As practitioners, we have to be practical and pragmatic. We have to try and do the best we can for this child, right here, right now.

That means that all views have to be accommodated and prioritised, and competing interests need to be balanced. No one individual can determine an outcome. Arguably the ideal situation is to reach a decision with the agreement of all concerned, but without letting the parents shoulder the burden of responsibility alone.

Although in theory there are one hundred and one reasons why agreement may be impossible, in practice it is (thankfully) not that common for serious problems or disagreements to arise around the best course of action to take. Unfortunately however, they do occur.

Tips on how to reach consensus and avoid disagreement in practice

In my experience, when problems and disagreements occur, it is usually down to blocked communication rather than blocked ethics. It is amazing how even enmeshed problems can untangle when everyone feels heard and feels able to hear. Remember, these two abilities are seriously strained when we are frightened and angry. Fear and anger are natural reactions to the death of a child, but not helpful ones. So we have to try and find a way through or around them. For example:

- Involve everyone whose view counts in the dilemma.
- Be completely honest and open with all, within the confines of each person's intellectual and emotional ability to understand.
- Be absolutely rigorous; go through all possibilities and scenarios with all the relevant people. Even the nasty ones.
- People make decisions with their heads and their hearts, so look for the hidden agendas as well as the expressed agendas.
- Be absolutely fair. Give each person a say, and be sure not to discriminate against anyone (especially the child).
- Be compassionate but be fair; empathise, try and put yourself in everyone's shoes, but try to avoid siding with those people whose emotional approach most closely mirrors your own.
- And, most importantly, **take time**: all of this takes time. Take as much as you can without being unfair on other patients or yourself.

What other resources might I find helpful?

- For further information regarding making decisions about withholding and withdrawing life-sustaining treatments (LST) see the Together for Short Lives's *Care Pathway for extubation within a children's palliative care framework* (add link)
- See the Royal College of Paediatrics and Child Health publication, Witholding *or Withdrawing Life Sustaining Treatment in Children: A Framework for Practice* (http://www.rcpch.ac.uk)
- Together for Short Lives, 2011. *A Parent's Guide: Making critical care choices for your child*. Bristol: Together for Short Lives.
- Together for Short Lives, 2011. *A Care Pathway to Support Extubation within a Children's Palliative Care Framework*. Bristol: Together for Short Lives. British Medical Association, 2007: *Withholding and withdrawing life prolonging medical treatment: guidance for decision making*. BMA: London.

- The Ethox Centre, 2010. *Ethical and Legal Issues at End of Life.* The Ethox Centre. http://www.ethox.org.uk/education/undergraduate-course/ethical-decision-making-at-the-end-of-life/End%20of%20Life%202010-11.doc/view
- General Medical Council (GMC), 2008. *Consent: Patients and Doctors Making Decisions Together.* London: GMC.
- Royal College of Paediatrics and Child Health (RCPCH), 2004. *Withholding or Withdrawing Life Saving Medical Treatment in Children: A framework for practice,* 2nd Edition. London: Royal College of Paediatrics and Child Health.
- South Central Strategic Health Authority (England), 2010. *Advance Care Plan policy* (extract from the Guide for Clinicians) (See http://www.oxfordshirepct.nhs.uk/about-us/documents/255Childandyoungpersonadvancedcareplanpolicyoctober2010.pdf)

Guidance
- End-of-life decisions - Views of the BMA can be found at: http://www.bma.org.uk/images/endlifedecisionsaug2009_tcm41-190116.pdf
- GMC: Treatment and care towards the end of life: good practice in decision making (pages 90-108) can be found at: http://www.gmc-uk.org/guidance/ethical_guidance/6858.asp
- BMA: End of life guidance can be found at: http://www.bma.org.uk/ethics/end_life_issues/

Toolkit
- The BMA: Mental Capacity Act Tool Kit. 2008 can be found at: http://www.bma.org.uk/ethics/consent_and_capacity/mencaptoolkit.jsp

Part eleven

How do I offer spiritual care to families?

What you probably know already

- Spiritual care is different from religious or pastoral care, but the two can be complementary.
- Spirituality has no standard definition and will be different for each person or family.
- Spirituality can be found at the heart of good palliative care.
- Pain can be measured, suffering cannot.

What you might find useful

Spirituality has been interpreted as whatever gives a person's life meaning. It may or may not include religion or a god. For reasons of clarity we would define spiritual and pastoral care as follows:

Spiritual care is responding to the uniqueness of the individual: accepting their range of doubts, beliefs and values just as they are. It means responding to the spoken or unspoken statements from the very core of that person as valid expressions of where they are and who they are. It involves being a facilitator in their search for identity on the journey of life and in the particular situation in which they find themselves. It involves responding without being prescriptive, judgmental or dogmatic and without preconditions, acknowledging that each will be at a different stage on that personal spiritual journey[105].

Pastoral care is the healing, sustaining, guiding, personal/societal formation and reconciling of persons and their relationships to family and community by a representative of their own faith (ordained or lay), and by their faith communities, who ground their care in the theological perspective of the faith tradition and who personally remain faithful to that faith through spiritual authenticity.

Spirituality is the proper concern of all who work with dying children and their families, but more so for those who have a role in direct patient care. Spirituality cannot be treated in isolation; it is firmly located within an overlapping and multi-layered context and needs to be considered from a number of perspectives.

Why is spiritual care so important in children's palliative care?

The child and family, and the team caring for the child all have spiritual needs and they all need to be included in spiritual care.

The task of offering spiritual care is that of co-creating a safe and secure or 'sacred' space, where the child and family can express their inner feelings or suffering and know that it is all right to do so, that they will be heard and taken seriously. It should also be a space where the parents, siblings and staff can have freedom to do the same. The child or family will not be able to do this work if their child is in any kind of pain.

When we think of 'pain' we need to think about the different kinds of pain. We may think the child is in physical pain and give medication, when in fact it could be a spiritual or soul pain. Soul pain is: *"the experience of an individual who has become disconnected and alienated from the deepest and most fundamental aspects of him or herself"*.

Soul pain is about 'intractable pain'. We need to try and understand the metaphors that people use to describe their distress, disease, their pain or their unconscious world. We have to learn to crack the code they use to tell us their story. We need to be very focused in our work when trying to assess and manage a child's pain. But sometimes when you are getting nowhere with a child who has 'intractable pain', you may need to switch your focus and take a wider view than the purely scientific, and think more about the experience of pain for the child.

Doctors and nurses are trained in communication, especially in the use of non-verbal communication and communicating with families. These skills transfer easily across into spiritual care. The art of good spiritual care is the art of *listening*. Being able to be still and to hear what is not being said. The skill is to be 'present', which means you should make sure you have time for the consultation, turning off your phone, not looking at your watch.

How do children develop their spirituality?

For children, spirituality is far more likely to be centred around their understanding of life as they experience it day-to-day.

Common spiritual concerns for children involve love, forgiveness, safety, hope and their legacy (will their life and their accomplishments have made a difference and will they be remembered after they have died).

Children are also likely to be concerned by loneliness and separation (from parents, siblings, pets, friends) and from loss of themselves as a whole (for example, no longer being able to go to school or do the things that they enjoy).

Children can sometimes talk about 'magical' and non-human beings such as angels, fairies or monsters. Listening to these stories, which may at first seem like childish storytelling may tell us something about a child's current 'spiritual' thinking and help us to better understand their fears and worries.

Some useful tips, phrases and activities to help spiritual dialogue with children

The most useful skill you can master is to listen, very deeply.
- **Listen to the words**: For example: God, heaven, spirit, hope, wish, anger, sad, ghost, lonely, strong, weak, guilty, brave or afraid. Explore these thoughts by asking what the words mean to the child.
- **Listen to the dreams:** The story or fears coming from dreams can give a chance to look at worries that are difficult to look at in 'real' life. Ask the child what the dream means to him/her. Do not try to explain it yourself.
- **Listen for 'searching' phrases:** Phrases that show the child is thinking or searching deeply can give you a chance to encourage the child to talk about it more. The child may ask *'why me?'* or *'I wish…'* or *'I wonder if'*. You can help the child work this out more by asking *'what else do you wish?'* or *'how do you think that may happen?'*
- **Listen to the journey:** Children who are beginning to sense that they are dying often talk about going home or leaving. Talking about these feelings and exploring the journey with the child is difficult but it needs to be done. Do not give false reassurance that they are not dying.

If you would like to prompt a child a little, here are some useful questions that might help get you started:
- What makes you feel safe?
- Who or what do you trust?
- How you know what is right and wrong?
- Who/what helps you when you need it?
- What do you make of what is happening to you?
- Who are you?
- What is important to you in life?
- What do you believe in?
- Do you believe in a god or spiritual being?
- What do you think happens to people when they die?

What other resources might I find helpful?

- Goldman, A. Hain, R. and Liben, S. (2006) *Oxford Textbook of Palliative Care for Children – Chapters 6, 7 and 16*. Oxford: Oxford University Press
- Pridmore, P. & Pridmore, J. (2004) 'Promoting the spiritual development of sick children', International Journal of Children's Spirituality, Vol. 9, No. 1.
- Editorial, 'Suffering and healing – our core business', Palliative Medicine 2009; 23: 385–387. http://pmj.sagepub.com/content/24/1/99.extract
- Ethnicity online: This site contains links to information about several ethnic / religious groups, including summaries of their beliefs and customs, along with healthcare-related advice: http://www.ethnicityonline.net/ethnic_groups.htm
- Interfaith Calendar: http://www.bbc.co.uk/religion/tools/calendar/
- Oxford Textbook of Palliative Care for Children. 2006. Ann Goldman, Richard Hain and Stephen Liben. Oxford. OUP
- 'Communicating with children' in 'Children's Palliative Care in Africa, 2009, Justin Amery (ED), Oxford, OUP. (downloadable for free at http://www.icpcn.org.uk/core/core_picker/download.asp?id=204))
- Pridmore, P. & J. Pridmore '(2004) 'Promoting the spiritual development of sick children', International Journal of Children's Spirituality, Vol. 9, No. 1,
- Editorial, 'Suffering and healing – our core business', Palliative Medicine 2009; 23: 385–387
- CASSELL, E, J. (1982) 'The Nature of Suffering & the Goals of Medicine' The New England Journal of Medicine, Vol 306, No 11, pages 639-645.
- KEARNEY, M. (1996). 'Mortally Wounded', Marino Books, Dublin.
- COLES, R. (1992) 'The Spiritual Life of Children', Harper Collins, London.
- SOMMER, D. R. (1989) 'The Spiritual Needs of Dying Children', Issues in Comprehensive Paediatric Nursing, Vol 12, pages 225-233

- STOTER, D. (1995) 'Spiritual Aspects of Health Care' Mosby, London.
- Cobb. M (2001) The Dying Soul: Spiritual Care at the End of Life Facing Death', Open University Press
- GOODLIFF. P (1998) Care in a confused climate: pastoral care and postmodern culture, Darton, Longman and Todd
- Himmelstein B, Hilden J, Boldt A, Weissman D: 'Pediatric Palliative Care' New England Journal of Medicine, Volume 350:1752-1762 April 22, 2004
- Chaplaincy network Birmingham Children's Hospital.

Part twelve

How do I provide good end of life care to a child and their family?

What you may know already

- Most children and adults prefer to die at home
- In palliative care, it is good to hope for the best but prepare for the worst
- Preparation is nine tenths of the issue
- Children, families and professionals are less anxious and more effective if they are prepared
- It is hard to talk about worst case scenarios in CPC

What you may not know

- Most children can be managed at home at the end of life
- There are a lot of resources and people to help you help your patients
- If you plan and discuss scenarios ahead, you can draw up an end-of-life symptom management plan so that everyone can be properly prepared
- Good end of life care requires good teamwork and planning
- For most of us, the idea of sitting down with a child and family and talking about the events leading up to the end of life, and beyond, and then scooping out possible worst case scenarios, is an extremely unwelcome prospect.

- But here is the thing: *it is almost impossible to provide good CPC without a good end-of-life care plan that everyone agrees to and that everyone buys in to*.
- Trust me, if you don't do it, there will almost certainly come a time where you or the family really regret it, but by then it will be too late.

What are the key goals for end of life care?

In the UK, the leading CPC charity 'Together for Short Lives' has identified some key goals for end of life care. They are probably valid anywhere in the world and they are:
- Professionals should be open and honest with families when the approach to end of life is recognised.
- Joint planning with families and relevant professionals should take place as soon as possible.
- The plan of care must consider local legislation with regard to medication in the community, resuscitation, and certification of death.
- A written plan of care should be agreed, which includes decisions about resuscitation, and emergency services should be informed.
- Care plans should be reviewed and altered to take account of changes.
- There should be 24 hour access to pain and symptom control, including access to medication.
- Those managing the control of symptoms should be suitably qualified and experienced wherever possible
- Emotional and spiritual support should be available to the child and family.
- Children and families should be supported in their choices and goals for quality of life to the end.

The Together for Short Lives 'Care Pathway' identifies three stages to end of life care:
- Recognition of end of life.
- Assessment of end of life needs and wishes.
- An end of life plan.

How do I recognise the end of life stage?

It may seem obvious, but the first, and possibly most important step in good end of life care is recognising that the child has probably reached the end of his or her life, and then 'naming' that for the child, family, carers, colleagues and yourself.

Unfortunately, determining the end of life stage for a child with a life-limiting condition can be difficult for a number of reasons:
- Children can be remarkably resilient and survive what we may think is 'the last event'. On the other hand, children can also decline very rapidly and unexpectedly. This is why it is ideal to have made some of the difficult end of life decisions beforehand, but also important to have parallel planning in place, where care prepares for the worst, but hopes for the best. This is especially important as there is considerable unpredictability of disease trajectory for many of these children.
- There are attitudinal barriers that make prognostication difficult, as none of us finds talking about a child's death easy, and some doctors and nurses can feel as if the death of a child is a personal failure [106].
- Sometimes the end of life is preceded by a period of aggressive efforts to save a child's life which may make it more difficult for the family to accept that their child has reached the end stage. The difficulty in accurately predicting death means families often face many acute life-threatening events thinking each one is a terminal event. This is unsurprisingly emotionally and physically exhausting for them.
- There could also be reluctance on the part of both parents and professionals to use certain drugs which often become necessary at the end of life for fear of 'causing death'.

On the other hand, in practice, it is usually not that difficult to guesstimate whether a child is likely to die within days, or within weeks, or within months. We know about the natural history of diseases, we can see children deteriorating and becoming weaker, and we may have access to investigations that tell us a terminal event is near.

Ultimately, there are no easy or reliable indicators of when the 'end of life' stage starts. Perhaps the most useful trigger for me has been to become aware as soon as the thought 'is this child nearing the end of life' pops into my head, I start planning. Trust your intuition. Don't ignore it!

How do I go about assessing and planning in practice?

Just as at any other time, the key to good end of life care is good assessment. The key tasks are to:

- **Assess the needs of the child and family**: Carefully assess the physical, psychological, family, social, spiritual and practical issues that might prevent a good death.
- **Identify the relevant decision makers**: Remember that if you don't get everyone on board, someone might derail even the best-laid plans at a crucial moment; perhaps when you are not around to set things straight.
- **Set the agenda for a meeting**: Make sure you identify all the issues early on, even if you don't cover them all in one go. It is important that all the decision makers understand what needs to be discussed, and make time to do it. The meeting doesn't have to be formal, but it does need to happen, sooner rather than later.
- **Meet and impart all the necessary information**: In order to plan effectively, all decision makers need to be in possession of all the relevant facts.
- **Agree which decisions need to be made by the child, which by the family, and which by the professionals.**
- **Get the decisions made**: If you think, for example, that one of your clinical colleagues might not want treatment withdrawn, or a child to be discharged, you don't have much time left to get this sorted out. Busy as you might be, you must be collaborative in planning, but also be prepared to be brave and decisive in order to ensure a 'good death'.

- **Talk about quality of life**: Explain how the child's quality of life might be adversely affected as death approaches and agree how you are going to manage these possibilities.
- **Draw up an end of life care plan**. Make sure the family and all relevant services have a copy of this. Think about withdrawal of life-sustaining treatment and any other unnecessary treatment: There may be drugs, artificial feeds or other treatments that are no longer necessary. Do the child and family want to continue these? Are they in the child's best interests? If so, you need to explain why and reach an agreed plan[107].
- **Implement the plan**: Get everything you need in place. Make sure that everyone has access to the relevant drugs and equipment. Check that everyone is in place and prepared, that the location is sorted out, and that everyone knows who is doing what.
- **Communicate**: The chances are that the plan you come up with will involve many people: the child, close and extended family, friends and carers, professionals and others. Does everyone know what they have to do and when? This is the area that most frequently goes wrong, so don't leave it to chance.
- **Plan for the worst, especially out of hours**: Give yourself a moment of peace to think about what could go wrong. Have you left any gaps? Is everyone clear? Do they have the necessary drugs and equipment? Who calls who if things go wrong?

How often should I assess a child at the end of life?

At the end of life, things usually speed up and events can change rapidly. Therefore, whereas you might have been assessing the child weekly, you may need to start assessing daily or even hourly. Robert Twycross[108] used to describe the 'rule of threes', based upon how quickly you think the patient is deteriorating:

- If a patient is deteriorating every day: assess every three days.
- If a patient is deteriorating every hour: assess every three hours.
- If a patient is deteriorating every minute: assess every three minutes.

How do I draw up an 'End of Life Care Plan'?
Ultimately, the aims of the end of life care plan are simple: to think through all of the child's and family's wishes for end of life care, to think through all the possible problems that may emerge, and to plan and prepare for each of them.

There are 2 main parts to an end-of-life care plan:
- A statement of wishes of the child and family regarding the care they wish for and hope for (sometimes called an 'Advanced Care Plan')
- A plan for managing medical eventualities that might arise as the child dies (sometimes called a 'Symptom Management Plan'). We will cover this in the next chapter.

The 'Statement of Wishes'.

The statement of wishes usually includes the child and families key thoughts and decisions. It should:
- Communicate the child's and family's wishes for the child's care, as part of the broader end of life care plan.
- Set out an agreed plan of care to be followed when critical events occur and/or when a child's condition deteriorates.
- Provide a framework for discussing and documenting the agreed wishes of a child and his/her parents, regarding specific care choices.
- Include decisions regarding the withdrawal of life-sustaining treatment which may include a DNACPR order.
- Guide professionals should parent(s) or next of kin cannot be contacted.

Below is a table taken from the 'Together for Short Lives' Care Pathway, listing things to think about when working with a family to draw up a statement of wishes.

Family
- Practical support
- Sibling involvement

- Grandparents
- Emotional support
- Spiritual/religious issues
- Cultural issues
- Funeral planning
- Organ donation

Child or young person
- Pain and symptom control
- Quality of life
- Friends
- Emotional support
- Spiritual/religious issues
- Cultural issues
- Funeral planning
- Organ donation
- Resuscitation/withdrawal of treatment (ACP/DNACPR)
- Special wishes or activities
- Life goals
- Children's 'will'
- Memory box

Environment
- Place of death
- Ambience
- Place of body after death

You could also use an adapted 'PEPSI COLA'[109] care plan, or have a look at an example from the UK[110].

When and how should I draw up a 'do not resuscitate' order?

The RCPCH has stipulated 5 situations where a DNACPR decision may be allowable[111]:

- **No chance situation**: The child has such severe disease that life-sustaining treatment simply delays death without significant alleviation of suffering. Treatment to sustain life is inappropriate.
- **No purpose situation**: Although the patient may be able to survive with treatment, the degree of physical or mental impairment will be so great that it is unreasonable to expect them to bear it.
- **Unbearable situation**: The child and/or family feel that in the face of progressive and irreversible illness further treatment is more than can be borne. They wish to have a particular treatment withdrawn or to refuse further treatment irrespective of the medical opinion that it may be of some benefit.
- **Permanent vegetative state.**
- **Brain stem death.**
- **The law of the country does not allow DNACPR**

If you think your patient may fit into one of these categories, it might be appropriate to draw up a 'do not attempt cardiopulmonary resuscitation (DNACPR)'. If so, you need to sit down with the family, and any other relevant professionals or carers, to discuss it. Key factors to discuss are:

- The DNACPR decision should reflect the agreed wishes of the child (where appropriate), those with parental responsibility for the child, and the professionals caring for the child. Such wishes may be a result of religious beliefs and will may require repeated discussions as the child's condition changes.
- The parents and child are not asked to sign the DNACPR as this would place an unnecessary stress on families to feel that they have to bear any responsibility for a DNACPR decision. It is for the lead professional to take ultimate responsibility.
- In situations where DNACPR is not legally permitted, and the child is in hospital, if available request a consult from the PICU/ICU team to discuss whether escalation of invasive resuscitation

would be appropriate. Such a decision must be discussed with the child and their parents and documented in the medical notes, without having to use the term DNR.
- Reasons for the DNACPR decision must be documented.
- The decision should be clearly recorded, signed and dated in the DNACPR section of the statement of wishes (or on a separate form according to local practice).
- It usually excludes reversible causes of arrest such as choking.
- For the form to be valid, the event should fall within the time period specified.
- Any important and relevant agencies such as other doctors and nurses involved with the family, hospitals, and ambulance services and so on must be told of the DNACPR decision as resuscitation is attempted on all children unless there is a valid DNACPR order in place.

In England, consideration should be given to notifying the local Child Death Overview Panel (CDOP) about the existence of a Statement of Wishes and DNACPR order. This should ensure that the local rapid response team is aware, should death occur suddenly/unexpectedly.

How do I prepare and plan for symptom management plan for the end of life?

Every child who is nearing the end of life should have an end-of-life symptom management plan.

To draw up a plan, you need to consider which symptoms the child might suffer from. This is usually not that difficult. Most children are at risk of pain, anxiety, and terminal agitation. If you think that opioids may need to be used, then there is a chance that the child with suffer nausea and vomiting. Often children will develop respiratory secretions (the 'death rattle'). If the child has clotting problems, has tumours that might invade blood vessels or has varices, you need to consider how you would manage bleeding. If the child has CNS disease, then you need to consider the possibility of seizures.

The process for drawing up a symptom-management plan is as follows
- Set sufficient time aside. This is not something you can do in ten minutes. It may take several meetings to agree and finalise.
- Consider the possible ways that the child might die, from most to least likely.
- List all of the symptoms that the child might suffer from.
- For each symptom write up what the child, family and carers might do non-pharmacologically
- For each symptom write up what the child, family and care team might do pharmacologically
- Sit down with the team and family and talk them through the plan, answer any questions, make any adjustments necessary to the plan, and ensure everyone knows what to do for each eventuality
- Put together a 'just in case' box containing all the drugs that might be needed according to your plan and ensure that it is left in the home for easy access
- Write up the plan, give a copy to the family, put another ion the just in case box and circulate it to any other agencies or professionals that might need to be involved.

'Just in Case Boxes'

Alder Hey Hospital has done some research[112] suggesting that the following six drugs can be considered essential
- Diamorphine;
- Cyclizine;
- Haloperidol;
- Levomepromazine;
- Midazolam;
- Hyoscine hydrobromide.
- Dexamethasone

You should also include any other equipment, dressings, catheters etc. may be needed for the child.

Symptom Management Plan Template

At the end of this book you will find a template for a symptom-control plan that you might find helpful. Please feel free to copy and use it.

What other resources might I find helpful?

- Together for Short Lives (2004) *A Framework for the Development of Integrated Multi-agency Care Pathways for Children with Life-threatening and Life-limiting Conditions.* Bristol: Together for Short Lives
- Amery, J. (Ed.) (2009) *Children's Palliative Care in Africa Chapter 18.* Oxford: Oxford University Press. Available from <http://www.icpcn.org.uk/core/core_picker/download.asp?id=204>
- Goldman, A. Hain, R. and Liben, S. (2006) *Oxford Textbook of Palliative Care for Children – Chapter 17.* Oxford: Oxford University Press
- Regnard, C. (Ed) and Dean, M. (Ed) (2010) *A guide to Symptom Relief in Palliative Care Revised edition (6th Edition)* Oxford: Radcliffe Publishers

<u>Guidance</u>
- South Central NHS (2010) Guide to Using the Child and Young Person's Advanced Care Plan [online] available from <Guide to Using Child and Young Persons ACP Policy>
- The Scottish Government (2011) *Resuscitation Planning Policy for Children and Young People (under 16 years)* [online] available from <http://www.scotland.gov.uk/Topics/Health/NHS-Scotland/LivingandDyingWell/CYPADM>

Part thirteen

How do I deal with the practicalities arising after the death of a child?

What you probably know already

- The rules for verification of death in your country.
- Children have varying degrees of capacity to contribute to decisions regarding end of life care (for example, DNACPR orders).
- Good inter-agency communication is essential both when planning for a child's end of life, and after death.

What you might find useful

- Each country has different regulations, procedures and customs in managing the practicalities of the death of a child
- Unless you deal with child death frequently, it is unlikely you will know all of them
- The more you find out ahead of time, the less likely you are to have an unwelcome surprise after
- Some of the questions you might want to ask yourself ahead of time might be:
    o Who do I need to inform?
    o Might there need to be a post-mortem?
    o Is organ-donation for transplant an option?

- o   What systems do I need to follow for registration of the death?
- o   What rituals will need to be followed after death?
- o   How quickly does the body need to be moved?
- o   Where will the body need to go, and how will it get there?
- o   Who will have to pay for this, and can they afford it?

What happens to the body after death?

This is a tricky chapter to write as regulations and rituals vary dramatically from one place to the next. There are some universals however.

Firstly, families need time with the body of their child after death if they are to grieve as healthily as possible. The less 'clinical' and 'institutional' you can make this, the better. So try and create some quiet room around the body, remove clinical paraphernalia, dress the child in home clothing with any treasured toys or objects, allow as much access to the family as possible, and give them as much control over preparing and managing the body of the child as possible.

Secondly, all bodies decay after death, so your ability to slow the process of decay down will have a big impact on what happens next. If you have access to a mortuary or cold room, the pressure is less acute. If you have access to undertakers, who can embalm the body, the pressure is less acute. But all this costs money. If you are working in a hot country, and the family have no access or can't afford these things, then you will need to move much faster, so it is better to be prepared and ready well before the actual death.

Rules and Regulations

All countries have systems for registering child deaths, and for investigating deaths that are unexplained, unexpected or suspicious. Sometimes, post-mortem testing is desirable to help identify the causes of death, especially where there may be a genetic component, and the family need help to assess future risks to other members. Also, in the world of CPC, while we don't like to think too much about child abuse, or malpractice, they do happen. Not every child death is entirely natural or unavoidable. So you need to find out early who to inform and how to go about it.

Organ donation

In many CPC scenarios, organ donation is not an option, because it is not available, or the dead child's organs are not in a fit state for transplantation. However, in some cases it is a possibility, and apart from the benefits to the person receiving the donated organ, it can be a powerful comfort to families to know that their child has an ongoing, living legacy. There are two types of organ donation; beating heart and non-beating heart.
- Beating heart donation is only considered in a child who has confirmed brain stem death.
- Non-beating heart donation is usually only considered for children who have a death that is expected within a specific time period, e.g. withdrawal of care, and can be any organ.

For most organs the child must be taken to theatre within ten minutes of death for organ harvesting. Corneas and heart valves can be harvested up to 48 hours after death. Tissue donation may be possible even when organ donation is not feasible. For more information contact your local transplant co-ordinator.

Rituals

Rituals are hugely important in helping us cope in times of transition, and there is arguably no more painful or traumatic transition than that of a child from life to death. Medical settings are rarely adequate to cope with the complexities and richness of these rituals, but we can at least try. It is tempting to think that you know the preferred rituals of a family you are caring for, especially if you come from a similar cultural background, but you would be surprised. When teaching in Uganda, I asked our group (all experienced doctors and nurses) to split into groups from different cultural backgrounds within Uganda. We had a few Brits in the group, so we got together too. Each group then presented what their particular tradition dictates. The variety was enormous, and nobody in the group was aware of all of the others. They included

- A period of lying in at home, during which mourners visit and impart their sympathies and support
- Moving the child to a mortuary
- Burying the child in the centre of the family home, sometimes bringing the dead body in through the door, and sometimes breaking through the wall
- Variation in the type of ceremony chosen dependent on the age and sex of the child
- Burying the child in earthenware pots, partly submerged in the lakes of the Northwest of the country
- Burying the child in a cemetery
- Cremating the child in crematorium
- Scattering ashes in a meaningful place
- Planting trees or other forms of remembrance

Each of these has powerful resonance in the culture within which it has developed, although arguably removing the body rapidly to a funeral director and thence to a crematorium (the most common practice of my own culture) has little cultural resonance at all, and has more to do with the denial and avoidance of mortality that characterises discourse about death and dying in the UK.

It can be hard even to recognise and understand all these various traditions, let alone enable them to happen in most medical settings, but it is amazing what can be done with a bit of preparation, forethought and creativity. These moments matter deeply, and are retained by the family forever, possibly more than any other time in a child's life.

So it is worth the effort.

What other resources might I find helpful?

Interesting article
- Davies, R. (2005) 'Mothers' stories of loss: their need to be with their dying child and their child's body after death (abstract).' *SAGE Journals Online* vol. 9 no. 4 288-300 [online] available from < http://chc.sagepub.com/content/9/4/288.abstract>
- National SIDS/Infant Death (2007) *Selected Resource for Grieving Parents, Their Families, Friends and Other Caregivers* [online] available from http://www.sidscenter.org/documents/SIDRC/BereavementSelectedResources.pdf
- UK Blood Transfusion and Tissue Transplantation Services (2011). Latest guidelines available from <http://www.transfusionguidelines.org.uk/Index.aspx?Publication=CTD&Section=17&pageid=1539>
- Ministry of Justice (2008) *Cremation Regulations Guidance for Doctors* [online] available from <http://www.justice.gov.uk/guidance/docs/cremation-doctors-guidance.pdf>

Part fourteen

How do I help the family with grief and bereavement?

What you probably know already

- Grief is the emotional and social reaction to loss, whereas bereavement is a state of having lost someone or something dear to you. Mourning is the external expression of loss.
- Families, communities and cultures may grieve and mourn differently. Rituals can help to bring healing and closure.
- It is important that health care professionals always respect the families' wishes and/or their cultural traditions and norms.
- Doctors and nurses can play a huge part in helping people grieve, simply by being a point of stability and a professional, experienced listener; as well as a gatekeeper to important practical support such as benefits, sickness certification and mental health support.

What you might find useful

- Grief is a natural consequence of the death of someone close, but when a child dies, the impact of their death is often much more distressing for the people close to them. The death of a child can be life-changing and family members, particularly parents, can grieve for a long time and may need ongoing bereavement support or counselling to help them move on with their life and cope with their loss.

- The death of a child often affects a large family group, including parents or carers, grandparents (who may be not only grieving for the child who has died, but also for the loss that their own child has experienced), siblings, cousins and extended or adoptive family.
- The dying child will also have had to come to terms with the anticipation of their own death.
- As a doctor or nurse you will have a lot of useful experience dealing with bereaved people, but it is important to keep in mind the heightened distress that the death of a child can cause, and also the differences in the way grief manifests itself in bereaved children and adults.

Bereaved parents

Bereaved parents need special attention. No-one can anticipate how they will feel after the death of their own child. Most parents describe a 'rollercoaster' of emotions, ranging from numbness to furious anger, profound sadness, to a certain relief. Seemingly irrational behaviour and reactions are also very common, as well as overwhelming physical exhaustion or compulsive activity.

Parents may want to talk to you just as a listening ear, but will often need to be referred to specialist services to deal with their loss, where these are available. Following the immediate loss of a child, it is important to maintain contact with the family, parents may feel bereft if they have required intense palliative support and suddenly it is withdrawn. It is helpful within the team to agree on a bereavement follow up plan to help families adjust to trying to resume a sense of normality to their lives

Some families may find it helpful to set aside an identified time each day within the first few months when they know they can focus on the death of their child, rather than feeling as if their grief consumes them every hour of the day. It can be helpful if difficult times such as birthdays, religious festivals or the anniversary of the child's death are remembered.

It can be useful to recommend the family create memory boxes and put some of the child's favourite things for example favourite blanket or teddy, anything which reminds them of their child in this special place. Memory books, including photographs can be a way for parents to release emotion and create objects which will remind them of their child.

It should be acknowledged that grief for a beloved child may never end or resolve, as Talbot, a bereaved mother and grief counsellor, notes: *"Healing after the death of a child does not mean becoming totally pain-free. Healing means integrating and learning how to live with the loss. It means being able to love others and reinvest in life again. Healing comes when parents decide that they will not permit pain to be the only expression of their continuing love for their child."* (Talbot, 2002)

Bereaved children

A dying child grieves for the impending loss of her own life, and of course siblings, other child relatives and friends of the dying child grieve too.

Sometimes it's difficult to talk to children about death in a way that they fully understand. Below are some tips relating to the unique situations where a child has lost a loved one who is also a child, including siblings, cousins and friends.

How do children grieve?

- It is difficult to offer sensible generalisations about how children react to death of different loved ones: their temperaments, personalities and circumstances are so varied.
- Children and young people may show a range of reactions and these are partly determined by their developmental stage.
- Their response will also vary according to the cause and nature of the death, the family circumstances, any previous experience of death or trauma within the family, the age and relationship with the person who has died, their position within the family,

- Children may have fluctuations in their grief between sudden sadness and equally suddenly appearing happy. This can be very confusing for you, and for them. Allow them to express these emotions, and support them through these feelings. Reassure them these feelings are 'normal' and they should be expressed.

How can I help a child through grief and bereavement?

A useful term is 'bereavement work'[113], which suggests that people have to work their way through various emotions, feelings, thoughts and behavioural effects within their bereavement experience.

From a carer's perspective, you should do what you can to try and help grieving children with their 'work', aiming to help them deal with it as effectively as possible.

Of course, the problem with this model is that no-one can say what 'effective' actually is for any one individual. However it may still be helpful to think in these terms, and look at the 'work' required in three stages:

1. **Pre-bereavement:** A child will begin to feel grief as soon as they understand that they or their loved one is going to die.
2. **At the time of death:** What happens at the time of death has profound implications for surviving loved ones. A painful, traumatic death will often leave survivors feeling guilty, angry and traumatised, whereas a 'good death' can help survivors to look back on the positives of a child's life.
3. **After death:** A child's grief can now focus on what has been lost (rather than what will be lost).

Pre-bereavement stage
- Encourage the child to talk and communicate, especially with family and friends.
- Avoid using abstract explanations such as *"your brother has gone to sleep"*.

- Allow the child to express emotions.
- Do not impose expectations on the child (e.g. by saying "*you will definitely feel better in time*").
- Encourage normality and continuity in other areas such as school.
- Allow denial if it occurs, but make it easy for the child to ask questions.
- Try to make the impending loss real for the child by including them in such activities as planning for burial ceremonies and last funeral rites.
- Children are very creative and may wish to participate in creating memory books to store special memories of their times together to look back on as they get older.
- Be prepared for anticipatory grief; often manifesting as separation anxiety, sadness, anger or withdrawal.
- Try to prevent the child becoming isolated if the family 'clams up' by explaining and encouraging more open discussion, and encouraging children to tell others.
- Remember young children think imaginatively, and may attribute huge consequences (including death) to tiny causes (e.g. that they upset mummy that day). Allow them to use play or story telling with dolls or teddy bears to enable them express their feelings, and to enable you to correct any false blame that they are taking.

After death
According to different cultures, it may or may not be the norm for children to see the body or attend the funeral. Seeing the body is often helpful, but should always be the child's choice.

Seeing the body may help a bereaved child to:
- Begin to say goodbye;
- Begin to accept the reality and finality of the death;
- Begin to understand what has happened
- Realise the child will not be coming back;
- Be less scared.

Before the child sees the body, give clear and detailed information about what will happen. For example, *"Joseph is lying on a bed. He doesn't look exactly the same as when he was alive. He is completely still He will not be able to talk and will not move. If you touch him he won't be warm, he may feel cold. He is wearing his pale shirt and his dark grey trousers. There are quite a lot of flowers in the room and also some cards."*

Let them choose what they do when they enter the room: keep still by the door, touch or stroke the body, leave something like a drawing with the body. Give them the choice as to whether they want someone with them, or whether they would like a little private time on their own.

Attending the funeral

- Try to ensure that they are with someone who will support them.
- Reassure them that it is all of the body of the person who has died that is being buried or cremated.
- Explain that the dead person can no longer feel anything or be scared.
- Explain that the dead person will no longer be suffering and is not in pain.
- Explain that there might be a 'party' after and not to be surprised or upset by that.
- Prepare them for some of the things that adults may say to them. For example, boys may be told that they are the 'man of the house now' and may appreciate reassurance that they are not.
- Create opportunities to be involved (e.g. through placing a drawing with the body or saying something at the funeral).

Alternative goodbyes

If the child cannot or does not want to attend the funeral, try to encourage an alternative goodbye ceremony to help with their grief.

Examples include holding a memorial ceremony at home or at the grave; visiting a place with special memories, creating a special place of their own choosing, releasing balloons with special messages, into the sky, lighting a candle and sharing special memories with each other, or starting a memory box or book.

What other resources might I find helpful?

- Bluebond-Langner, M. (1980) *The Private Worlds Of Dying Children*, Princeton: Princeton University Press.
- Goldman, A. Hain, R. and Liben, S. (2006) *Oxford Textbook of Palliative Care for Children – Chapter 15*. Oxford: Oxford University Press

Telephone and Internet support
- The Child Death Helpline: The Child Death Helpline is a helpline for anyone affected by the death of a child of any age, from pre-birth to adult, under any circumstances, however recently or long ago: http://www.childdeathhelpline.org/ Helpline: 0800 282 986 Email: contact@childdeathhelpline.org
- Winston's Wish is a childhood bereavement charity that provides services to bereaved children, young people and their families: http://www.winstonswish.org.uk/
- The Compassionate Friends UK is an organisation of bereaved parents and their families offering understanding, support and encouragement to others after the death of a child or children. They also offer support, advice and information to other relatives, friends and professionals who are helping the family: www.tcf.org.uk Helpline: 0845 123 2304 Email: info@tcf.org.uk
- TCF Sibling Support is a project run by The Compassionate Friends which provides nationwide self-help support for people who have suffered the loss of a brother or sister. www.tcfsiblingsupport.org.uk
- The Child Bereavement Charity http://www.childbereavement.org.uk

- Grief Encounter aims to help and support each person with an individual http://www.griefencounter.org.uk
- Season for Growth is a loss and grief peer-group education programme for young people aged 6-18 years in England & Wales: http://seasonsforgrowth.co.uk/ and for Scotland: http://www.notredamecentre.org.uk/seasons-for-growth.aspx.htm

Part fifteen

How do I survive and thrive in children's palliative care?

What you probably know already

- Being a health professional is both rewarding and challenging.
- Whether you thrive with the rewards or buckle under the challenges depends on numerous factors, many (but by no means all) of which are under your control.
- You can make positive or negative choices, and you can act constructively or destructively.

What you might find useful

- Being a professional involves self-understanding, recognition of strengths and weaknesses, ability to pace oneself, mastering the tools of the trade, and keeping those tools sharp.
- Burnout is less common in palliative care than in similar professions.[114,115,116,117].
- Children's palliative care creates opportunities to be challenged, work within a good team, and feel like we are doing something that has meaning and purpose. These are all key ingredients for happiness, which means there is no theoretical reason why we should not be happy in our work, even though it contains great sadness
- Nevertheless the overall prevalence of mental health problems in palliative care physicians is 25% which, although no worse than other doctors and nurses and students, is not exactly good.

What factors might make me stressed or burnt out?

Over-stress and burnout are not just damaging to the individual, but also to the team and patients, as they cause irritability, paranoia, argumentativeness, slow working, mistakes, resentment, poor communication, loss of empathy and patience, and eventual sickness.

There are numerous factors that the literature suggests may lead to stress and burn out:

Personal factors
- Feeling out of control/out of one's depth.
- Feeling unsupported.
- Personal loneliness.
- Caring for others at home.

Psychological factors
- Mental health problems.
- Substance misuse.
- Unresolved personal trauma.

Organisational factors
- Work overload.
- Lack of role-clarity/work-life boundaries.
- Resource constraints.
- Fear of job loss, discipline, bullying or other abuse at work.
- Too much change.
- Unrealistic goals.

Team factors
- A team providing palliative care fundamentally exists to contain pain and grief within itself. When the level of pain outweighs the resilience of the team, it will start to split [118], and this is quite common in children's palliative care. [119].
- Team splitting can manifest as:
    - Scapegoating: Team-members demonise an individual and project all their negative emotions onto him or her.

- Sub-group (or clique) formation: Where the team splits into different sub-groups, each with different agendas and values.
- Psychological 'splitting' of the team: A bit like scapegoating, but involving projection of negative emotions onto sub-groups rather than individuals (i.e. where one subgroup demonises another subgroup).
- Change-avoidance: Where team members stick rigidly to the familiar, even where improvements are needed.
- Team burn-out: Which shows itself as poor morale, poor quality of care, chronic in-fighting and team divisions.

Patient factors
- Perhaps counter-intuitively, coping with death and dying do not emerge as a major source of job stress among children's palliative care professionals.[120].
- However, there are certain patient factors which are more likely to overwhelm your defences:
  - When the patient is young.
  - When the patient reminds you of someone close to you, or something/someone from your past.
  - When the death is traumatic.
  - When you have formed a close relationship with the patient.
  - When several deaths occur in a short space of time.

What can make me more resilient?

If you have spotted several risk factors for burnout that apply to you, you may be feeling worried. But remember to be optimistic. If you are still turning up for work for more than simply to pick up the pay cheque, you must have tremendous powers of resilience. What's more, your resilience can be built up in a number of ways:

Psychological strengthening
The increased incidence of past psycho-social trauma in health workers can actually be more of a strength than a weakness. It means we will have

learnt through experience how to be empathic, understanding and able to communicate with people who are sick or dying.

For it to be a strength rather than a weakness, psychological work needs to be done to acknowledge the pain and accept when you cannot 'save' someone. That means valuing yourself enough to set boundaries that are strong enough for personal protection, and flexible enough for when patient needs and work circumstances change. That psychological work involves:

Personal work
- Health and energy: Eating, sleeping and exercising well.
- Optimism: We can't really change the world, but we can try to see the best rather than the worst in things.
- Common sense: Seeing big problems merely as lots of little ones small enough to tackle.

Social skills work
- Standing up for yourself: There is more suffering and there are more patients in the world than anyone can possibly help with. You have to be able to say when you've done enough for one day.
- Practice and experience: The highest risk of burnout tends to occur in the first two to three years of a new job, thereafter it declines.
- Having fun: What do you enjoy doing? Are you doing it? If not, why not?
- Overwork: If you are working too hard, stop it. If you don't absolutely have to for financial reasons, ease up. If you consistently push yourself over the limits, your performance could start to suffer.
- Get connected: You might be isolated. You may not have a broad circle of friends. However hard it might be, try and nurture relationships with others. Other people can make you feel appreciated, they can help put things in perspective, and they can offer someone to talk to.

Organisational and team work
- Remember the importance of team work in order to provide good care.
- Clarify your objectives: Make sure you have a proper appraisal or review. It's important to review where you are and plan where you are going.
- Get adequate supervision: Supervision provides a system for talking through difficult cases, sharing problems, sifting solutions and planning the way forward.
- Team support meetings: Different colleagues will be at different stages, so those who are flying can support those who are struggling. As a result, the team will develop a sense of group responsibility and trust for each other.
- If these things aren't happening, speak to your boss. If you are the boss, explain to yourself why these are important.

What do I do if I think I might be burning out?

Feeling that you might be burning out or suffering psychological illness such as anxiety, depression or substance addiction can be extremely frightening. When we are frightened at this existential level our psychological defences often kick in, sending us into denial or subverting the problem. However, one effective way of dealing with these problems is to look at your symptoms as friends rather than enemies.

Our burn-out, our anxiety, our low mood, and our addictive behaviour are all helpful messages from our subcsonsciousness to our consciousness. It is telling us that things are not right, that we are under threat, and that we must do something to protect ourselves.

'Doing something' when we are down is tricky, but not impossible. We can use the same skills and capabilities with ourselves that we use with our patients: assess, measure, analyse, plan, and act. For example we can:
- Assess yourself – ask yourself honestly: Where am I? What burnout factors are at play?

- Do a personal audit - Try running a burnout inventory, or an anxiety, depression or substance misuse score on yourself. Or talk it over with a friend, colleague or your own health carer. Believe the results. Note them down so you can check progress in the future.
- Analyse your audits - Which parts of your life are stressing you the most? Think broadly: personal factors, personality type, team factors, organisational factors and environmental factors.
- Dare to dream - Ask yourself this question: What do I want to have happen? Allow yourself to suspend reality temporarily. If you had a magic wand, how would you change your life and what would it look like?
- Set goals and objectives – Come back to reality. Even if all of your dream appears unachievable, there will definitely be steps that you can take to set off in the right direction. What are they? Write down at least three achievable goals for 1 year, 1 month, 1 week and 1 day. You can do this alone, but it is usually easier and more effective to talk it through with someone else who can 'coach' you through the process.
- Implement: This bit might be particularly hard if you are a bit burnt out (remember resistance to change is an early feature). But no-one is going to change your life for you. You need to act if you are to change. So choose at least one of the objectives that you have set yourself to do today, and do it. It may be small but even the longest and hardest journey begins with that first step. By moving, however hesitantly and slowly, you are sending yourself a message: I am in the wrong place, I need to get to the right place, and I am worth the effort of moving. That positive reinforcement is powerful, and will give you more energy to take a second step
- Be gentle with yourself and smile at your imperfections: Every journey includes mis-steps, backward steps and false steps. When you go wrong, drop back or fall over, don't give up. We learn much more from our mistakes than from our successes. Pick yourself up, brush yourself down, smile at your own weaknesses and congratulate yourself on your efforts – just as you would with at a child learning to ride a bike. Eventually, you will stay up.

And remember, when we are suffering from anxiety, depression or substance abuse, it is very difficult to pull ourselves back just by carrying on and hoping for the best. Trying to do everything alone is a perverse egoic behaviour which denies ourselves the love and support that the universe always offers to us for free. We are there for others, so others will be there for us.

Remember, even if you worked every hour of every day, you would hardly scratch the surface of human suffering. So be honest and real with yourself. You owe it to your patients, your colleagues, your family and yourself to get healthy and stay healthy.

Ask for help.

Justin Amery

Start close in,
don't take the second step
or the third,
start with the first
thing
close in,
the step
you don't want to take.

Start with
the ground
you know,
the pale ground
beneath your feet,
your own
way of starting
the conversation.

Start with your own
question,
give up on other
people's questions,
don't let them
smother something
simple.

To find
another's voice,
follow
your own voice,
wait until
that voice
becomes a
private ear
listening
to another.

Start right now
take a small step
you can call your own
don't follow
someone else's
heroics, be humble
and focused,
start close in,
don't mistake
that other
for your own.

Start close in,
don't take
the second step
or the third,
start with the first
thing
close in,
the step
you don't want to take.

~David Whyte, River Flow: New and Selected Poems

Do you think you are a bit burnt out? If so, try a burnout inventory -http://www.mindtools.com/stress/Brn/BurnoutSelfTest.htm
Keep a Stress Diary http://www.stress-management-for-peak-performance.com/stress-diary.html
Check out if you are depressed or anxious http://discoveryhealth.queendom.com/depression_abridged_access.html; http://www.goodmedicine.org.uk/files/general%20anxiety,%20assessment%20gadss,%20tahoma.DOC
Are you taking too many substances? CAGE Questionnaire http://counsellingresource.com/quizzes/alcohol-cage/index.html

What other resources might I find helpful?

Amery, J. Surviving and Thriving in Health Practice. Oxford. Radcliffe Publishing. 2013.

Amery, J. (Ed.) (2009) *Children's Palliative Care in Africa*. Oxford: Oxford University Press. Available from <http://www.icpcn.org.uk/core/core_picker/download.asp?id=204> - Chapter 19
BMA Counselling/ Doctors for Doctors: http://www.bma.org.uk/doctors_health/index.jsp

Here are three websites, which may be of help:
Narcotics Anonymous: http://www.ukna.org/
Alcoholics Anonymous: http://www.aa.org/
Cocaine Anonymous: http://www.cauk.org.uk/

The British Doctors' and Dentists' Group: For drug and alcohol users: (http://www.bddg.org/page.php?id=1)
The British International Doctors Association: Where cultural or linguistic problems may be a contributing factor, doctors can access the health counselling panel: Tel: 0161 456 7828; Email: oda@doctors.org.uk
Doctors' support network: Self-help group for doctors with any form of mental health concern. They also have a confidential, anonymous peer support telephone line: http://www.dsn.org.uk/

Medical Careers: There are a range of mental health conditions that lead to a disability. These include anxiety, bipolar disorder, depression, obsessive-compulsive disorder and schizophrenia. Mental health conditions are often associated with alcohol and drug abuse, as well as eating disorders. Many people acquire mental illnesses during their working life. http://www.medicalcareers.nhs.uk/career_options/doctors_with_disabilities/doctors_with_mental_health_con.aspx

The Association of Paediatric Palliative Medicine Master Formulary
3rd edition

2015

Introduction

Welcome to the third edition of the APPM formulary. Even in the short time between the publications of the two editions there have been some major changes in the use of certain medications. Many of the drugs have been extensively rewritten and references have been brought up to date. The publication of the WHO guide on pain 2012 is closely reflected in this document with the changes in the WHO ladder for pain.
We have decided that rather than produce lengthy monographs of each drug we would instead focus on key practice points pertaining to individual drugs. We have focused on use in palliative care and only included this specific use and excluded the better known and more general indications the view being that other information would be easily obtainable from other national formularies. We have included a note about the licensing status for each drug.

For each individual drug, evidence is cited from research papers (where available) on its usage. We have also cited the source(s) used for where drug dosages have been obtained. In many cases the evidence for use of some drugs has been either weak or extrapolated from adult dosages. In some situations dosage is based on clinical consensus. Although this is not necessarily the best way to give drugs to children we have been mindful of the fact that research of drug usage in children and specifically in children's palliative care is difficult, and as yet still in its infancy in this small but rapidly developing field.
We have included only those drugs, routes and indications generally used in children's palliative care in Great Britain. The drugs are presented here in alphabetical order by generic name. We would strongly advise practitioners not to prescribe outside their expertise, and if in doubt to consult the growing network of clinicians with specialist expertise in paediatric palliative medicine. For some drugs, higher doses than noted here may be recommended by specialists in the field familiar with their use.

Justin Amery

We hope that over the course of time our colleagues around the world will communicate to us ways in which we can improve this formulary. Please do let us know of any omissions or additions that you feel we should add to the formulary by e-mailing appm@togetherforshortlives.org.uk.

It is hoped that other formularies in books or hospitals will base their information on this master formulary in the field of paediatric palliative medicine. All the key paediatric palliative formularies used around the UK have already agreed to adopt the style and content of this master formulary.

This formulary is provided free of charge and all the contributors work to improve paediatric palliative care around the world. Feel free to make as many copies as you like but please do not alter, plagiarise or try to copy any of the work into your own name. If you wish to use the work in a specific way then contact us for approval (sat.jassal@gmail.com).

Abbreviations

RE = strong research evidence
SR = some weak research evidence
CC = no published evidence but has clinical consensus
EA = evidence (research or clinical consensus) with adults
SC = subcutaneous
IV = intravenous
IM = intramuscular
CSCI = continuous subcutaneous infusion

In general (and when available), this formulary includes, for palliative care, the same doses as those recommended in one or more of: British National Formulary (BNF)[1], British National Formulary for Children (BNFC)[2], Neonatal Formulary[3], WHO guidelines on the pharmacological treatment of persisting pain in children with medical illnesses[4], Palliative Care Formulary[5] and Medicines for Children[6]. Readers outside the UK are advised to consult any local prescribing guidelines in addition to this Formulary..

The authors have made every effort to check current data sheets and literature up to September 2014, but the dosages, indications, contraindications and adverse effects of drugs change over time as new information is obtained. It is the responsibility of the prescriber to check this information with the manufacturer's current data sheet and we strongly urge the reader to do this before administering any of the drugs in this document. In addition, palliative care uses a number of drugs for indications or by routes that are not licensed by the manufacturer. In the UK such unlicensed use is allowed, but at the discretion and with the responsibility of the prescriber.

Copyright protected
© APPM.

Formulary

Adrenaline (topical)

Use:
> Small external bleeds.

Dose and routes:
> Soak gauze in 1:1000 (1 mg/mL) solution and apply directly to bleeding point.

Evidence: [1] CC

Alfentanil

Use:
- Short acting synthetic lipophilic opioid analgesic derivative of fentanyl.
- Used as analgesic especially intra-operatively and for patients in intensive care and on assisted ventilation (adjunct to anaesthesia).
- Alternative opioid if intolerant to other strong opioids; useful in renal failure if neurotoxic on morphine, or stage 4 to 5 severe renal failure.
- Useful for breakthrough pain and procedure-related pain.

Dose and Routes:

Analgesic especially intra-operatively and for patients in intensive care and on assisted ventilation (adjunct to anaesthesia). **SEEK SPECIALIST ADVICE**

By IV/SC bolus *(**these doses assume** assisted ventilation is available)*
- **Neonate:** 5-20 micrograms/kg initial dose, supplemental doses up to 10 micrograms/kg,
- **1 month to 18 years:** 10-20 micrograms/kg initial dose, up to 10 micrograms/kg supplemental doses.

By continuous IV or SC infusion *(**these doses assume** assisted ventilation is available)*
- **Neonate:** 10-50 micrograms/kg over 10 minutes then 30-60 micrograms /kg/ hour,
- **1 month to 18 years**: 50-100 microgram/kg loading dose over 10 minutes, then 30-60microgram/kg/hour as a continuous infusion

Alternative opioid if intolerant to other strong opioids; useful in renal failure if neurotoxic on morphine, or stage 4 to 5 severe renal failure. **SEEK SPECIALIST ADVICE**

Doses should be based on opioid equivalence with the following suggested as safe and practical conversion ratios
Oral morphine to CSCI alfentanil: $1/30^{th}$ of the 24 hour total oral morphine dose e.g. 60mg/24hours oral morphine = 2mg/24 hours CSCI alfentanil

CSCI/IV morphine to CSCI alfentanil: $1/15^{th}$ of the 24 hour total CSCI/IV morphine dose e.g. 30mg/24hours CSCI/IV morphine = 2mg/24 hours CSCI alfentanil

CSCI diamorphine to CSCI alfentanil: $1/10^{th}$ of the 24 hour total diamorphine dose e.g. 30mg/24 hours diamorphine = 3mg/24 hours CSCI alfentanil

If conversion is due to toxicity of the previous opioid, lower doses of alfentanil may be needed to provide adequate analgesia.
Opioid naive Adults: CSCI 500microgram-1mg over 24 hours

Breakthrough pain SEEK SPECIALIST ADVICE

SC / Sublingual / Buccal

Suggest 1/6th to 1/10th of the total CSCI dose. However there is a poor relationship between the effective PRN dose and the regular background dose. Alfentanil has a short duration of action (~30 minutes) and even with an optimally titrated PRN dose, frequent dosing (even every 1-2 hours) may be required. Dose and frequency of administration should be regularly reviewed.

Procedure-related pain SEEK SPECIALIST ADVICE

SC / Sublingual / Buccal
- **Adults:** 250-500microgram single dose
- **Child:** 5microgram/kg single dose

Give dose 5 minutes before an event likely to cause pain; repeat if needed

Notes:
- Alfentanil injection is licensed for use in children as an analgesic supplement for use before and during anaesthesia. Use for pain relief in palliative care is unlicensed. Buccal, sublingual or intranasal administration of alfentanil for incident/breakthrough pain is an unlicensed indication and route of administration. The injection solution may be used for buccal, sublingual or intranasal administration (unlicensed).
- There is limited information / evidence for analgesic doses in palliative care, especially in children. Doses are largely extrapolated from suggested equianalgesic doses with other opioids.
- Potency: 10-20 times stronger than parenteral morphine, approximately 25% of the potency of fentanyl.
- Very useful in patients with severe renal failure (no dose reduction is needed). May need to reduce the dose in severe hepatic impairment.

- In order to avoid excessive dosage in obese children, the dose may need to be calculated on the basis of ideal weight for height, rather than actual weight.
- Pharmacokinetics: half-life is prolonged in neonates, so can accumulate in prolonged use. Clearance may be increased from 1 month to 12 years of age, so higher infusion doses may be needed.
- Contraindication: not to be administered concurrently with MAOIs (monoamine oxidase inhibitors) or within 2 weeks of their discontinuation.
- Interaction: alfentanil levels are increased by inhibitors of Cytochrome P450.
- Adverse effects include respiratory depression, hypotension, hypothermia, muscle rigidity (which can be managed with neuromuscular blocking drugs).
- For SC or IV infusion, alfentanil is compatible with 0.9% NaCl or 5% glucose as a diluent. For CSCI alfentanil appears compatible with most drugs used in a syringe driver. Like diamorphine, high doses of alfentanil may be dissolved in small volumes of diluent which is very useful for SC administration.
- Available as: injection (500 microgram/mL in 2ml and 10ml ampoule); Intensive care injection (5 mg/mL in 1ml ampoule which must be diluted before use). Nasal spray with attachment for buccal / SL use (5 mg/5 mL bottle available as special order from Torbay Hospital: each 'spray' delivers 0.14ml = 140 microgram alfentanil).

Evidence: [1, 5-9]

EA, RE (for PICU settings), CC (in palliative care settings outside ICU)

Amitriptyline

Use:
- Neuropathic pain.

Dose and routes:
By mouth:
- **Child 2–12 years:** initial dose of 200microgram/kg (maximum 10 mg) given once daily at night. Dose may be increased gradually, if necessary, to a suggested maximum of 1mg/kg/dose twice daily (under specialist supervision).
- **Child 12–18 years:** initial dose of 10 mg at night increased gradually, if necessary, every 3-5 days to a suggested initial maximum of 75 mg/day. Higher doses up to 150 mg/day in divided doses may be used under specialist advice.

Notes:
- Not licensed for use in children with neuropathic pain.
- Analgesic effect unlikely to be evident for several days. Potential improved sleep and appetite which are likely to precede analgesic effect.
- Drug interactions: not to be administered concurrently with MAOIs (monoamine oxidase inhibitors) or within 2 weeks of their discontinuation. Caution with concurrent use of drugs which inhibit or induce CYP2D6 enzymes.
- Main side effects limiting use in children include; constipation, dry mouth and drowsiness.
- Liquid may be administered via an enteral feeding tube.
- Available as: tablets (10 mg, 25 mg, 50 mg) and oral solution (25 mg/5 mL, 50mg/5mL).

Evidence: [1, 10-12]

Aprepitant

Use:
- Prevention and treatment of nausea and vomiting associated with emetogenic cancer chemotherapy

Dose and route:
For oral administration:
- **Child** <10 years: 3mg/kg (max 125mg) as a single dose on Day 1 (1 hour before chemotherapy) followed by 2mg/kg (max 80mg) as a single dose on Day 2 and Day 3
- **Child** >10 years: 125mg as a single dose on Day 1 (1 hour before chemotherapy) followed by 80mg as a single dose on Day 2 and Day 3

Aprepitant is used in combination with a corticosteroid (usually dexamethasone) and a 5-HT3 antagonist such as ondansetron.

Notes:
- Aprepitant is licensed for the prevention of acute and delayed nausea and vomiting associated with highly or moderately emetogenic cancer chemotherapy in adults
- Aprepitant is not licensed for use in children and adolescents less than 18 years of age (although a number of clinical trials are currently ongoing). Limited evidence for use in those less than 10 years of age
- Aprepitant is a selective high-affinity antagonist at NK_1 receptors
- Aprepitant is a substrate, a moderate inhibitor and inducer of the CYP3A4 isoenzyme system. It is also an inducer of CYP2C9 and therefore has the potential to interact with any other drugs that are also metabolised by these enzyme systems including rifampicin, carbamazepine, phenobarbital, itraconazole, clarithromycin, warfarin and dexamethasone. Please note this list is not exhaustive – seek advice.

- Common side effects include hiccups, dyspepsia, diarrhoea, constipation, anorexia, asthenia, headache and dizziness
- Available as: capsules 80mg and 125mg. A formulation for extemporaneous preparation of an oral suspension is available.

Evidence: [1, 5, 13-17]

Arachis Oil Enema

Use:
- Faecal softener
- Faecal impaction

Dose and route:
By rectal administration
- **Child 3-7 years:** 45-65 mL as required (~1/3 to 1/2 enema),
- **Child 7-12 years:** 65 mL - 100 mL as required (~1/2 to 3/4 enema),
- **Child 12 years and over:** 100-130 mL as required (~3/4 – 1 enema).

Notes:
- Caution: as arachis oil is derived from peanuts, do not use in children with a known allergy to peanuts.
- Generally used as a retention enema to soften impacted faeces. May be instilled and left overnight to soften the stool.
- Warm enema before use by placing in warm water.
- Administration may cause local irritation.
- Licensed for use in children from 3 years of age.
- Available as: enema, arachis (peanut) oil in 130 mL single dose disposable packs.

Evidence: [1, 6] CC

Aspirin

Use:
- Mild to moderate pain.
- Pyrexia.

Dose and routes:
By mouth:
- **> 16 years of age:** Initial dose of 300 mg every 4–6 hours when necessary. Dose may be increased if necessary to a maximum of 900 mg every 4-6 hours (maximum 4 g/day).

Notes:
- Contraindicated in children due to risk of Reye Syndrome.
- Use with caution in asthma, previous peptic ulceration, severe hepatic or renal impairment.
- May be used in low dose under specialist advice for children with some cardiac conditions.
- Available as: tablets (75 mg, 300 mg), dispersible tablets (75 mg, 300 mg), and suppositories (150 mg available from special-order manufacturers or specialist importing companies).

Evidence: [1]

Baclofen

Use:
- Chronic severe spasticity of voluntary muscle
- Considered as third line neuropathic agent

Dose and routes:
By mouth:
- **Initial dose for child under 18 years:** 300microgram/kg/day in 4 divided doses (maximum single dose 2.5 mg) increased gradually at weekly intervals to a usual maintenance dose of 0.75-2 mg/kg/day in divided doses with the following maximum daily doses:

- **Child up to 8 years:** maximum total daily dose 40 mg/day,
- **Child 8-18 years:** maximum total daily dose 60 mg/day,

Notes:
- Review treatment if no benefit within 6 weeks of achieving maximum dose.
- There is very limited clinical data on the use of baclofen in children under the age of one year. Use in this patient population should be based on the physician's consideration of individual benefit and risk of therapy.
- Monitor and review reduction in muscle tone and potential adverse effects on swallow and airway protection.
- Avoid abrupt withdrawal.
- Intrathecal use by specialist only.
- Risk of toxicity in renal impairment; use smaller oral doses and increase dosage interval if necessary.
- Contraindicated if there is a history of active peptic ulceration.
- Administration with or after food may minimise gastric irritation.
- May be administered via enteral feeding tubes. Use liquid formulation for small doses, dilute prior to use to reduce viscosity. Consider dispersing tablets in water for higher doses owing to the sorbitol content of the liquid formulation.
- Available as: tablets (10 mg) and oral solution (5 mg/5 mL).

Evidence: [1, 2, 12, 18-25]

Bethanechol

Use:
- Opioid induced urinary retention

Dose and routes:
By mouth:
- **Child over 1 year:** 0.6 mg/kg/day in 3 or 4 divided doses. Maximum single dose 10 mg.
- **Adult dose:** 10-25 mg per dose 3 to 4 times a day. Occasionally it may be felt necessary to initiate therapy with a 50mg dose.

Subcutaneous:
- **Child over 1 year:** 0.12 to 2 mg/kg/day in 3 or 4 divided doses. Maximum single dose 2.5 mg,
- **Adult dose:** 2.5 to 5 mg per dose 3 to 4 times a day.

Notes
- The safety and efficacy of bethanechol in children has not been established (bethanechol is not licensed for use in children).
- Preferably taken before food to reduce potential for nausea and vomiting.
- Contraindicated in hyperthyroidism, peptic ulcer, asthma, cardiac disease and epilepsy.
- Tablets may be crushed and dispersed in water for administration via an enteral feeding tube; formulation for extemporaneous oral suspension is available.
- Available as: tablets (10 mg and 25 mg), injection for subcutaneous injection only (5 mg/mL – not licensed in the UK but may be possible to import via a specialist importation company).

Evidence: [12, 26, 27]

Bisacodyl

Use:
- Constipation

Dose and routes:
By mouth:
- **Child 4–18 years:** 5-20mg once daily; adjust according to response.

By rectum (suppository):
- **Child 2–18 years:** 5-10 mg once daily; adjust according to response.

Notes:
- Tablets act in 10–12 hours. Suppositories act in 20–60 min; suppositories must be in direct contact with mucosal wall.
- Stimulant laxative.
- Prolonged or excessive use can cause electrolyte disturbance.
- Available as: tablets (5 mg) and suppositories (5 mg, 10 mg).

Evidence: [1, 2]

Buprenorphine

Use:
- Moderate to severe pain

Dose and routes:
By sublingual route (starting doses):
- **Child body weight 16–25 kg:** 100 microgram every 6–8 hours,
- **Child body weight 25–37.5 kg:** 100–200 microgram every 6–8 hours,
- **Child body weight 37.5–50 kg:** 200–300 microgram every 6–8 hours,
- **Child body weight over 50 kg:** 200–400 microgram every 6–8 hours.

By transdermal patch:
- By titration or as indicated by existing opioid needs.

Buprenorphine patches are *approximately* equivalent to the following 24-hour doses of oral morphine

morphine salt 12 mg daily	≡ *BuTrans®* '5' patch	7-day patches
morphine salt 24 mg daily	≡ *BuTrans®* '10' patch	7-day patches
morphine salt 48 mg daily	≡ *BuTrans®* '20' patch	7-day patches
morphine salt 84 mg daily	≡ *Transtec®* '35' patch	4-day patches
morphine salt 126 mg daily	≡ *Transtec®* '52.5' patch	4-day patches
morphine salt 168 mg daily	≡ *Transtec®* '70' patch	4-day patches

Notes:
- Sublingual tablets not licensed for use in children < 6 years old.
- Patches not licensed for use in children.
- Has both opioid agonist and antagonist properties and may precipitate withdrawal symptoms, including pain, in children dependant on high doses of other opioids.
- Sublingual duration of action 6-8 hours.
- Caution with hepatic impairment and potential interaction with many drugs including anti-retrovirals.
- Available as: tablets (200 microgram, 400 microgram) for sublingual administration. Tablets may be halved.
- Available as: two types of patches:
    1. BuTrans®—applied every 7 days. Available as 5 (5 microgram /hour for 7 days),10 (10 microgram /hour for 7 days), and 20 (20 microgram /hour for 7 days)
    2. TransTec®—applied every 96 hours. Available as 35 (35 microgram /hour for 96 hours), 52.5 (52.5 microgram /hour for 96 hours), and 70 (70 microgram /hour for 96 hours).

For patches, systemic analgesic concentrations are generally reached within 12–24 hours but levels continue to rise for 32–54 hours. If converting from:
- 4-hourly oral morphine - give regular doses for the first 12 hours after applying the patch

- - 12-hourly slow release morphine - apply the patch and give the final slow release dose at the same time
  - 24-hourly slow release morphine - apply the patch 12 hours after the final slow release dose
  - Continuous subcutaneous infusion - continue the syringe driver for about 12hours after applying the patch.
- Effects only partially reversed by naloxone.
- Rate of absorption from patch is affected by temperature, so caution with pyrexia or increased external temperature such as hot baths: possibility of accidental overdose with respiratory depression.
- Patches are finding a use as an easily administered option for low dose background opioid analgesia in a stable situation, for example in severe neurological impairment.
- Schedule 3 CD

Evidence: [2, 5, 28-30]

Carbamazepine

Use:
- Neuropathic pain.
- Some movement disorders.
- Anticonvulsant

Dose and routes
By mouth:
- **Child 1 month–12 years:** initial dose of 5 mg/kg at night or 2.5 mg/kg twice daily, increased as necessary by 2.5–5 mg/kg every 3–7 days; usual maintenance dose 5 mg/kg 2–3 times daily. Doses up to 20 mg/kg/day in divided doses have been used.
- **Child 12–18 years:** initial dose of 100–200 mg 1–2 times daily; increased slowly to usual maintenance of 200-400 mg 2–3 times daily. Maximum 1.8 g/day in divided doses.

By rectum:
- **Child 1 month–18 years:** use approximately 25% more than the oral dose (maximum single dose 250 mg) up to 4 times a day.

Notes:
- Not licensed for use in children with neuropathic pain.
- Can cause serious blood, hepatic, and skin disorders. Parents should be taught how to recognise signs of these conditions, particularly leucopoenia.
- Numerous interactions with other drugs including chemotherapy drugs.
- Different preparations may vary in bioavailability so avoid changing formulations or brands.
- Suppositories of 125 mg are approximately equivalent to 100 mg tablets.
- Oral liquid has been administered rectally – should be retained for at least 2 hours if possible but may have a laxative effect
- For administration via an enteral feeding tube use the liquid preparation. Dilute with an equal volume of water immediately prior to administration. If giving doses higher than 400 mg/day, divide into 4 equal doses.
- Available as: tablets (100 mg, 200 mg, 400 mg), chew tabs (100 mg, 200 mg), liquid (100 mg/5mL), suppositories (125 mg, 250 mg), and modified release tablets (200mg, 400 mg).

Evidence: [2, 12, 31-34]

Celecoxib

Use:
- Pain, inflammatory pain, bone pain, stiffness. Not used first line
- Dose based on management of juvenile rheumatoid arthritis

Dose and routes
By mouth:
- **Child over 2 years:**
  - Weight 10-25 kg: 50 mg twice daily or 100mg daily
  - Weight more than 25 kg: 100 mg twice daily

Notes
- Celecoxib is a cyclooxygenase-2 selective inhibitor
- Not licensed in the UK for use in children
- All NSAID use (including cyclo-oxygenase-2 selective inhibitors) can, to varying degrees, be associated with a small increased risk of thrombotic events (e.g. myocardial infarction and stroke) independent of baseline cardiovascular risk factors or duration of NSAID use; however, the greatest risk may be in those receiving high doses long term. COX-2 inhibitors are associated with an increased risk of thrombotic effects.
- All NSAIDs are associated with serious gastro-intestinal toxicity. COX-2 inhibitors are associated with a *lower risk* of serious upper gastro-intestinal side-effects than non-selective NSAIDs.
- Use with caution in patients with renal impairment and avoid in severe renal impairment.
- Use with caution in hepatic impairment.
- Celecoxib interacts with a great many commonly used drugs, check BNF (current version on-line).
- Capsules may be opened and contents mixed with soft food immediately before administration. For a 50mg dose, approximately halve the 100mg capsule contents to give a best estimate of a 50mg dose.
- Available as: capsules 100mg, 200mg.

Evidence: [1, 35-38] SR

Chloral hydrate

Use:
- Insomnia.
- Agitation

Dose and routes:
By mouth or rectum:
- **Neonate:** initial dose of 30 mg/kg as a single dose at night. May be increased to 45 mg/kg at night or when required,

- **Child 1 month–12 years:** initial dose of 30 mg/kg as a single dose at night. May be increased to 50 mg/kg at night or when required. Maximum single dose 1 g,
- **Child 12–18 years:** initial dose of 500 mg as a single dose at night or when required. Dose may be increased if necessary to 1-2 g. Maximum single dose 2 g.

Notes:
- Not licensed in agitation or in infants <2 years for insomnia
- Oral use: mix with plenty of juice, water, or milk to reduce gastric irritation and disguise the unpleasant taste.
- For rectal administration use oral solution or suppositories (available from 'specials' manufacturers).
- Accumulates on prolonged use and should be avoided in severe renal or hepatic impairment.
- Available as: tablets (chloral betaine 707 mg = choral hydrate 414 mg— Welldorm®), oral solution (143.3 mg/5 mL—Welldorm®; 200 mg/5 mL, 500 mg/5 mL both of which are available from 'specials' manufacturers or specialist importing companies), suppositories (available as various strengths 25mg, 50 mg, 60 mg, 100 mg, 200 mg, 500 mg from 'specials' manufacturers).

Evidence: [2, 3, 6, 39-41]

Chlorpromazine

Use:
- Hiccups
- Nausea and vomiting of terminal illness (where other drugs are unsuitable)

Dose and routes:
Hiccups
By mouth:
- **Child 1–6 years:** 500 micrograms/kg every 4–6 hours adjusted according to response (maximum 40 mg daily),

- **Child 6–12 years:** 10 mg 3 times daily, adjusted according to response (maximum 75 mg daily),
- **Child 12–18 years:** 25 mg 3 times daily (*or* 75 mg at night), adjusted according to response, higher doses may be used by specialist units.

Nausea and vomiting of terminal illness (where other drugs are unsuitable)
By mouth:
- **Child 1–6 years:** 500 micrograms/kg every 4–6 hours; maximum 40 mg daily,
- **Child 6–12 years**: 500 micrograms/kg every 4–6 hours; maximum 75 mg daily,
- **Child 12–18 years:** 10–25 mg every 4–6 hours.

By deep intramuscular injection:
- **Child 1–6 years:** 500 micrograms/kg every 6–8 hours; maximum 40 mg daily,
- **Child 6–12 years**: 500 micrograms/kg every 6–8 hours; maximum 75 mg daily,
- **Child 12–18 years:** initially 25 mg then 25–50 mg every 3–4 hours until vomiting stops.

Notes:
- Not licensed in children for intractable hiccup
- Caution in children with hepatic impairment (can precipitate coma), renal impairment (start with small dose; increased cerebral sensitivity), cardiovascular disease, epilepsy (and conditions predisposing to epilepsy), depression, myasthenia gravis.
- Caution is also required in severe respiratory disease and in children with a history of jaundice or who have blood dyscrasias (perform blood counts if unexplained infection or fever develops).
- Photosensitisation may occur with higher dosages; children should avoid direct sunlight.
- Antipsychotic drugs may be contra-indicated in CNS depression.

- Risk of contact sensitisation; tablets should not be crushed and solution should be handled with care.
- Available as: tablets, coated (25 mg, 50 mg, 100 mg); oral solution (25 mg/5 mL, 100 mg/ 5mL); injection (25 mg/mL in 1mL and 2mL ampoules).

Evidence: [1, 2, 42-50]

Clobazam

Uses:
- Adjunctive therapy for epilepsy

Dose and route:
For oral administration:
- **Child 1month - 6 years:** initial dose of 125 microgram/kg twice daily. Increase every 5 days as necessary and as tolerated to a usual maintenance dose of 250 microgram/kg twice daily. Maximum dose 500 microgram/kg (15 mg single dose) twice daily,
- **Child 6-18 years:** initial dose of 5 mg daily. Increase every 5 days as necessary and as tolerated to a usual maintenance dose of 0.3-1 mg/kg daily. Maximum 60 mg daily. Daily doses of up to 30 mg may be given as a single dose at bedtime, higher doses should be divided.

Notes:
- Not licensed for use in children less than 6 years of age.
- Once titrated to an effective dose of clobazam, patients should remain on their treatment and care should be exercised when changing between different formulations.
- Tablets can be administered whole, or crushed and mixed in apple sauce. The 10mg tablets can be divided into equal halves of 5mg. Clobazam can be given with or without food.
- Possible side-effects as would be expected from benzodiazepines. Children are more susceptible to sedation and paradoxical emotional reactions.

- Available as: tablets (10mg Frisium[(R)]); tablets (5mg – unlicensed and available on a named-patient basis); oral liquid (5mg in 5ml and 10mg in 5ml – care with differing strengths).
- Frisium[(R)] tablets are NHS black-listed except for epilepsy and endorsed 'SLS'.

Evidence: [2, 6]

Clonazepam

Use:
- Tonic-clonic seizures
- Partial seizures
- Cluster seizures
- Myoclonus
- Status epilepticus (3rd line, particularly in neonates)
- Neuropathic pain
- Restless legs
- Gasping
- Anxiety and panic

Dose and routes:
By mouth *(anticonvulsant doses: reduce for other indications)*:
- **Child 1 month–1 year:** initially 250micrograms at night for 4 nights, increased over 2–4 weeks to usual maintenance dose of 0.5–1 mg at night (may be given in 3 divided doses if necessary),
- **Child 1–5 years:** initially 250micrograms at night for 4 nights, increased over 2–4 weeks to usual maintenance of 1–3 mg at night (may be given in 3 divided doses if necessary),
- **Child 5–12 years:** initially 500micrograms at night for 4 nights, increased over 2–4 weeks to usual maintenance dose of 3–6 mg at night (may be given in 3 divided doses if necessary),
- **Child 12–18 years:** initially 1 mg at night for 4 nights, increased over 2–4 weeks to usual maintenance of 4–8 mg at night (may be given in 3 divided doses if necessary).

For status epilepticus: (SR)

Continuous subcutaneous Infusion:
- **Child 1 month – 18 years:** starting dose 20 - 25 microgram/kg/24 hours,
- Maximum starting doses: 1-5 years: 250 microgram/24 hours; 5-12 years: 500 microgram/24 hours
- Increase at intervals of not less than 12 hours to 200 microgram/kg/24 hours (maximum 8 mg/24 hours);
- Doses of up to 1.4 mg/kg/24 hours have been used in status epilepticus in PICU environment.

By intravenous injection over at least 2 minutes, or infusion:
- **Neonate:** 100 microgram / kg intravenous over at least 2 minutes, repeated after 24 hours if necessary (avoid unless no safer alternative). Used for seizures not controlled with phenobarbital or phenytoin,
- **Child 1 month to 12 years:** loading dose 50 micrograms/kg (maximum 1 mg) by IV injection followed by IV infusion of 10 microgram/kg/hour adjusted according to response; maximum 60 micrograms/kg/hour,
- **Child 12-18 years:** loading dose 1 mg by IV injection followed by IV infusion of 10 microgram/kg/hour adjusted according to response; maximum 60 micrograms/kg/hour.

Notes
- Licensed for use in children for status epilepticus and epilepsy. Not licensed for neuropathic pain. Tablets licensed in children.
- Very effective anticonvulsant, usually 3rd line due to side effects and development of tolerance.
- Use lower doses for panic, anxiolysis, terminal sedation, neuropathic pain, and restless legs.
- Do not use in acute or severe respiratory insufficiency unless in the imminently dying. Be cautious in those with chronic respiratory disease

- As an anxiolytic / sedative clonazepam is approximately 20 times as potent as diazepam (i.e. 250 microgram clonazepam equivalent to 5 mg diazepam orally).
- Multiple indications in addition to anticonvulsant activity can make clonazepam particularly useful in the palliative care of children for neurological disorders.
- Many children with complex seizure disorders are on twice daily doses and on higher dosages.
- The dose may be increased for short periods 3-5 days with increased seizures e.g. from viral illness
- Elimination half life of 20 - 40 hours means that it may take up to 6 days to reach steady state; there is a risk of accumulation and toxicity with rapid increase of infusion; consider loading dose to reach steady state more quickly.
- Avoid abrupt withdrawal
- Associated with salivary hypersecretion and drooling
- Tablets may be dispersed in water for administration via an enteral feeding tube.
- Stability of diluted clonazepam is up to 12 hours so prescribers should consider 12 hourly infusions.
- Compatible with most drugs commonly administered via continuous subcutaneous infusion via syringe driver.
- Available as: tablets (500 microgram scored, 2 mg scored); liquid (0.5 mg in 5 mL and 2 mg in 5 mL now available as licensed preparations from Rosemont but not indicated in children due to high alcohol content; other unlicensed oral liquids are available from specials manufacturers); injection (1 mg/mL unlicensed).

Evidence: [2, 3, 24, 33, 51, 52]

Clonidine

Uses:
- Anxiety / sedation (prior to procedure)
- Pain / sedation / opioid sparing / prevention of opioid withdrawal effects

- Regional nerve block
- Spasticity
- Behavioural symptoms of irritability, impulsiveness, aggression

Doses and routes:

Anxiety / Sedation / Pre-procedure:
Oral / Intranasal /Rectal:
- **Child >1 month:** 4 microgram/kg as a single dose (suggested maximum 150 microgram single dose). If used as premedicant prior to a procedure give 45-60 minutes before.

Pain / Sedation / Opioid sparing / Prevention of opioid withdrawal effects (most experience on PICU):
Oral / IV Bolus:
- **Child >1 month:** initial dose 1 microgram/kg/dose 3-4 times daily. Increase gradually as needed and tolerated to maximum of 5 microgram/kg/dose four times a day

IV infusion:
- **Child >1 month:** 0.1-2 microgram/kg/hour.
  Usual starting doses:
  - **Child <6 months:** 0.4 microgram/kg/hour;
  - **Child >6 months:** 0.6 microgram/kg/hour

For chronic long-term pain, and once an effective oral dose has been established, consideration can be made to transferring to transdermal patches using a patch size that will give a roughly equivalent daily dose of clonidine (see notes below).

Regional nerve block – only in situations where specialist input is available:
- **Child >3 months:** 1-2 microgram/kg clonidine in combination with a local anaesthetic

Spasticity / Movement Disorder:
Oral:
- **Child > 1 month:** 1-5 microgram/kg/dose three times a day. Frequency of dosing may need to be increased and /or alternative route of administration considered if the enteral route is not possible.

Behavioural problems / Tics / Tourette's syndrome:
Oral:
- **Child > 4 years:** Oral: initial dose of 25 microgram at night. Increase as necessary after 1-2 weeks to 50 microgram at night. Dose can be further increased by 25 microgram every 2 weeks to suggested maximum of 5 microgram/kg/day or 300 microgram/day

Notes
- Clonidine is a mixed alpha-1 and alpha-2 agonist (mainly alpha-2). Appears to have synergistic analgesic effects with opioids and prevent opioid withdrawal symptoms. Also useful for its sedative effect. Use established in ADHD, behavioural problems and tics.
- Not licensed for use in children
- Licensed indication of clonidine is for the treatment of hypertension so reduction in BP is a likely side effect of use. Titrate the dose of clonidine against the symptoms and monitor BP and pulse on starting treatment and after each dose increase.
- When used for longer than a few days, clonidine should be withdrawn slowly on discontinuation to prevent acute withdrawal symptoms including rebound hypertension.
- Use with caution in those with bradyarrhythmia, Raynaud's or other occlusive peripheral vascular disease.
- Common side effects include constipation, nausea, dry mouth, vomiting, postural hypotension, dizziness, sleep disturbances, headache.
- Effects of clonidine are abolished by drugs with alpha-2 antagonistic activity e.g. tricyclics and antipsychotic drugs.

- Antihypertensive effects may be potentiated by other drugs used to lower BP.
- Oral bioavailability 75-100%; generally 1:1 conversion IV: oral is suggested as a starting point (largely adult data; note: it has been suggested that oral bioavailability may be lower in children [53]).
- Some reports of use of rectal clonidine. Pharmacokinetic studies suggest almost 100% bioavailability via this route. Single rectal doses of 2.5-4microgram/kg have been used.
- Onset of effect: oral 30-60 mins. Time to peak plasma concentration: oral 1.5-5 hours; epidural 20 minutes; transdermal 2 days.
- Clonidine has been used successfully by SC injection and infusion – seek specialist advice.
- Oral solution may be administered via an enteral feeding tube. Alternatively, if the required dose is appropriate to the available tablet strengths, the tablets may be crushed and dispersed in water for administration via an enteral feeding tube. Note: the 25microgram tablets do not appear to disperse in water as readily as the 100microgram tablets.
- Chronic conditions – for older children the use of transdermal patches may be considered when an effective oral dose has been established which is great enough to allow an approximate conversion (1:1) to the transdermal route. Although unlicensed in the UK, 3 strengths of patches are available programmed to release 100 microgram, 200 microgram or 300 microgram clonidine daily for 7 days.
- Available as: tablets 25 microgram, 100 microgram; injection 150 micrograms/ml; transdermal patch 100 microgram, 200 microgram or 300 microgram clonidine daily for 7 days (not licensed in UK – available via importation company); oral solution (special) 50 microgram/ml

Evidence: [53-67]

Co-danthramer (dantron and poloxamer 188)

Use:
- Constipation in terminal illness only

Dose and routes:
By mouth:
Co-danthramer 25/200 suspension 5mL = one co-danthramer 25/200 capsule (Dantron 25 mg poloxamer '188' 200mg):
- **Child 2–12 years:** 2.5–5 mL at night,
- **Child 6–12 years:** 1 capsule at night,
- **Child 12–18 years:** 5–10 mL or 1–2 capsules at night. Dosage can be increased up to 10-20 mL twice a day.

Strong co-danthramer 75/1000 suspension 5 mL = two strong co-danthramer 37.5/500 capsules:
- **Child 12–18 years:** 5 mL or 1–2 capsules at night.

Notes
- Co-danthramer is made from dantron and poloxamer '188'.
- Acts as a stimulant laxative.
- Avoid prolonged skin contact due to risk of irritation and excoriation (avoid in urinary or faecal incontinence/ children with nappies).
- Dantron can turn urine red/brown.
- Rodent studies indicate potential carcinogenic risk.

Evidence: [1, 2]

Co-danthrusate (Dantron and Docusate Sodium)

Use:
- Constipation in terminal illness only

Dose and routes:

By mouth:

Co-danthrusate 50/60 suspension 5 mL = one co-danthrusate 50/60 capsule (Dantron 50mg/ Docusate sodium 60mg)
- **Child 6–12 years:** 5 mL or 1 capsule at night,
- **Child 12–18 years:** 5–15 mL or 1–3 capsules at night.

Notes
- Co-danthrusate is made from dantron and docusate sodium.
- Acts as a stimulant laxative.
- Avoid prolonged skin contact due to risk of irritation and excoriation (avoid in urinary or faecal incontinence/ children with nappies).
- Danthron can turn urine red/brown.
- Rodent studies indicate potential carcinogenic risk.

Evidence: [1, 2]

Codeine Phosphate

The European Medicines Agency's Pharmacovigilance Risk Assessment Committee (PRAC) has addressed safety concerns with codeine-containing medicines when used for the management of pain in children June 2013. This follows the PRAC's review of reports of children who developed serious adverse effects or died after taking codeine for pain relief. Children who are 'ultra rapid metabolisers' of codeine are at risk of severe opioid toxicity due to rapid and uncontrolled conversion of codeine into morphine.

The PRAC recommended the following risk-minimisation measures to ensure that only children for whom benefits are greater than the risks are given the medicine for pain relief:
- Codeine-containing medicines should only be used to treat acute (short lived) moderate pain in children above 12 years of age, and only if it cannot be relieved by other analgesics such

as paracetamol or ibuprofen, because of the risk of respiratory depression associated with codeine use.
- Codeine should not be used at all in children (aged below 18 years) with known obstructive airway disease or those who undergo surgery for the removal of the tonsils or adenoids to treat obstructive sleep apnoea, as these patients are more susceptible to respiratory problems.

http://www.ema.europa.eu/ema/index.jsp?curl=pages/news_and_events/news/2013/06/news_detail_001813.jsp&mid=WC0b01ac058004d5c1

Further, the WHO now advises there is insufficient evidence to make a recommendation for an alternative to codeine and recommends moving directly from non-opioids (Step 1) to low dose strong opioids for the management of moderate uncontrolled pain in children.

Uses:
- Mild to moderate pain in patients who through previous use are known to be able to benefit when other agents are contraindicated or not appropriate. For when required use only – not suitable for management of background pain.
- Marked diarrhoea, when other agents are contra-indicated or not appropriate, with medication doses and interval titrated to effect
- Cough suppressant

Dose and routes:
By mouth, rectum, SC injection, or by IM injection:
- **Neonate:** 0.5–1 mg/kg every 4–6 hours,
- **Child 1 month–12 years:** 0.5–1 mg/kg every 4–6 hours; maximum 240 mg daily,
- **Child 12–18 years:** 30–60 mg every 4–6 hours; maximum 240 mg daily.

As cough suppressant in the form of pholcodine linctus/syrup (NB/ Different strengths are available)
- **Child 6-12 years:** 2.5 mg 3-4 times daily,
- **Child 12-18 years:** 5-10 mg 3-4 times daily.

Notes:
- Not licensed for use in children < 1 year old.
- Codeine is effectively a pro drug for morphine, delivering approximately 1 mg of morphine for every 10 mg of codeine.
- Pharmacologically, codeine is no different from morphine except that it is weaker and less consistently effective. This has led the WHO to recommend that it is better replaced by low doses of morphine.
- Conversion to morphine is subject to wide pharmacogenetic variation. 5-34% of population have an enzyme deficiency that prevents activation of codeine to active metabolite and so it is ineffective in this group.
- Individuals who are ultra-rapid metabolisers can develop life threatening opioid toxicity.
- Seems relatively constipating compared with morphine/diamorphine, particularly in children.
- Rectal administration is an unlicensed route of administration using an unlicensed product.
- Must *not* be given IV.
- Reduce dose in renal impairment.
- Available as: tablets (15 mg, 30 mg, 60 mg), oral solution (25 mg/5 mL), injection (60 mg/mL), suppositories of various strengths available from 'specials' manufacturers. Pholcodine as linctus 2 mg/5 mL, 5 mg/5 mL and 10 mg/5 mL.
- Some retail pharmacies do not stock codeine phosphate solution at 25 mg/5 mL. They usually do stock codeine phosphate linctus at 15 mg/5 mL and this is worth enquiring of if a practitioner is working in the community and wishes to prescribe this medication. BE CAREFUL WITH DIFFERING STRENGTHS OF LIQUIDS.

Evidence: [1-3, 33, 68, 69]

Cyclizine

Use:
- Antiemetic of choice for raised intracranial pressure.
- Nausea and vomiting where other more specific antiemetics (metoclopramide, 5HT3 antagonists) have failed

Dose and routes:
By mouth or by slow IV injection over 3–5min:
- **Child 1 month–6 years:** 0.5–1 mg/kg up to 3 times daily; maximum single dose 25 mg,
- **Child 6–12 years:** 25 mg up to 3 times daily,
- **Child 12–18 years:** 50 mg up to 3 times daily.

By rectum:
- **Child 2–6 years:** 12.5 mg up to 3 times daily,
- **Child 6–12 years:** 25 mg up to 3 times daily,
- **Child 12–18 years:** 50 mg up to 3 times daily.

By continuous IV or SC infusion:
- **Child 1 month–5 years:** 3 mg/kg over 24 hours (maximum 50 mg/24 hours),
- **Child 6–12 years:** 75 mg over 24 hours,
- **Child 12–18 years:** 150 mg over 24 hours.

Notes:
- Antihistaminic antimuscarinic antiemetic
- Tablets are not licensed for use in children < 6 years old.
- Injection is not licensed for use in children.
- Antimuscarinic side effects include dry mouth; drowsiness, headache, fatigue, dizziness, thickening of bronchial secretions, nervousness.
- Rapid SC or IV bolus can lead to 'lightheadedness' –disliked by some and enthralling to others leading to repeated quests for IV Cyclizine.
- Care with subcutaneous or intravenous infusion – acidic pH and can cause injection site reactions

- For CSCI or IV infusion, dilute only with water for injection or 5% dextrose; *incompatible* with 0.9 % saline and will precipitate.
- Concentration dependant incompatibility with alfentanil, dexamethasone, diamorphine and oxycodone
- Suppositories must be kept refrigerated.
- Tablets may be crushed for oral administration. The tablets do not disperse well in water but if shaken in 10ml water for 5 minutes; the resulting dispersion may be administered immediately via an enteral feeding tube.
- Available as: tablets (50 mg), suppositories (12.5 mg, 25 mg, 50 mg, 100 mg from 'specials' manufacturers) and injection (50 mg/mL).

Evidence: [2, 12, 70]

Dantrolene

Use:
- Skeletal muscle relaxant.
- Chronic severe muscle spasm or spasticity.

Dose and routes:
The dose of dantrolene should be built up slowly
By mouth:
- **Child 5–12 years:** initial dose of 500 microgram/kg once daily; after 7 days increase to 500 microgram/kg/dose 3 times daily. Every 7 days increase by a further 500 microgram/kg/dose until response. Maximum recommended dose is 2 mg/kg 3–4 times daily (maximum total daily dose 400 mg),
- **Child 12–18 years:** initial dose of 25 mg once daily; after 7 days increase to 25 mg 3 times daily. Every 7 days increase by a further 500 microgram/kg/dose until response. Maximum recommended dose is 2 mg/kg 3–4 times daily (maximum total daily dose 400 mg).

Notes:
- Not licensed for use in children.

- Hepatotoxicity risk, consider checking liver function before and at regular intervals during therapy. Contraindicated in hepatic impairment: avoid in liver disease or concomitant use of hepatotoxic drugs.
- Can cause drowsiness, dizziness, weakness and diarrhoea
- Available as: capsules (25 mg, 100 mg), oral suspension (extemporaneous formulation).

Evidence: [2, 19, 20, 25, 71, 72]

Dexamethasone

Use
- Headache associated with raised intracranial pressure caused by a tumour.
- Anti-inflammatory in brain and other tumours causing pressure on nerves, bone or obstruction of hollow viscus.
- Analgesic role in nerve compression, spinal cord compression and bone pain.
- Antiemetic either as an adjuvant or in highly emetogenic cytotoxic therapies.

Dose and routes
Prescribe as dexamethasone base

Headache associated with raised intracranial pressure
By mouth or IV:
Child 1 month–12 years: 250 microgram/kg twice a day for 5 days; then reduce or stop.

To relieve symptoms of brain or other tumour
Numerous other indications in cancer management such as spinal cord and/or nerve compression, some causes of dyspnoea, bone pain, superior vena caval obstruction etc, only in discussion with specialist palliative medicine team. High doses < 16 mg/ 24 hrs may be advised.

Antiemetic

By mouth or IV:
- **Child < 1 year:** initial dose 250microgram 3 times daily. This dose may be increased as necessary and as tolerated up to 1 mg 3 times daily,
- **Child 1–5 years:** initial dose 1 mg 3 times daily. This dose may be increased as necessary and as tolerated up to 2 mg 3 times daily,
- **Child 6–12 years:** initial dose 2 mg 3 times daily. This dose may be increased as necessary and as tolerated up to 4 mg 3 times daily,
- **Child 12–18 years:** 4 mg 3 times daily.

Notes:
- Not licensed for use in children as an antiemetic.
- Dexamethasone has high glucocorticoid activity but insignificant mineralocorticoid activity so is particularly suited for high dose anti-inflammatory therapy.
- Dexamethasone can be given in a single daily dose each morning for most indications; this reduces the likelihood of corticosteroid induced insomnia and agitation
- Dexamethasone has an oral bioavailability of >80%; it can be converted to SC or IV on a 1:1 basis
- Dexamethasone 1 mg = dexamethasone phosphate 1.2 mg = dexamethasone sodium phosphate 1.3 mg.
- Dexamethasone 1 mg = 7 mg prednisolone (anti-inflammatory equivalence).
- Dexamethasone has a long duration of action
- Problems of weight gain and Cushingoid appearance are major problems specifically in children. All specialist units therefore use pulsed dose regimes in preference to continual use. Regimes vary with conditions and specialist units. Seek local specialist advice.
- Other side effects include: diabetes, osteoporosis, muscle wasting, peptic ulceration and behavioural problems particularly agitation.
- Dexamethasone can be stopped abruptly if given for a short duration of time (<7days), otherwise gradual withdrawal is advised

- Tablets may be dispersed in water if oral liquid unavailable. Oral solution or tablets dispersed in water may be administered via an enteral feeding tube.
- Available as: tablets (500 microgram, 2 mg), oral solution (2 mg/5 mL and other strengths available from 'specials' manufacturers) and injection as dexamethasone sodium phosphate (equivalent to 4 mg/1 mL dexamethasone base (Organon[R] brand) or 3.3 mg/mL dexamethasone base (Hospira[R] brand).

Evidence: [6, 49, 73-76]

Diamorphine

Use:
- Moderate to severe pain
- Dyspnoea

Dose and routes:
Normally convert using oral morphine equivalent (OME) from previous analgesia.
Use the following **starting** doses in opioid naive patient. The maximum dose stated applies to **starting** dose only.

Acute or Chronic pain

By continuous subcutaneous or intravenous infusion
- **Neonate:** Initial dose of 2.5 microgram/kg/hour which can be increased as necessary to a suggested maximum of 7 micrograms/kg/hour,
- **Child 1 month-18 years:** 7-25 microgram/kg/hour (initial maximum 10 mg/24 hours) adjusted according to response.

By IV /SC or IM injection:
- **Neonate:** 15 micrograms/kg every 6 hours as necessary, adjusted according to response,

- **Child 1-3 month:** 20 micrograms/kg every 6 hours as necessary, adjusted according to response,
- **Child 3-6 months:** 25-50 micrograms/kg every 6 hours as necessary, adjusted according to response,
- **Child 6-12 months:** 75 micrograms/kg every 4 hours as necessary, adjusted according to response,
- **Child 1-12 years:** 75-100 micrograms/kg every 4 hours as necessary, adjusted according to response. Suggested initial maximum dose of 2.5mg,
- **Child 12-18 years:** 75-100 micrograms/kg every 4 hours as necessary, adjusted according to response. Suggested initial maximum dose of 2.5-5mg.

By intranasal or buccal route:
- **Child over 10kg:** 50-100 micrograms/kg; maximum single dose 10 mg. Injection solution can be used by intranasal or buccal routes or Nasal spray (Ayendi(R)) now available and licensed for use in children aged 2 years and over (weight 12kg upwards) for the management of severe acute pain.

720microgram/actuation
- 12-18kg: 2 sprays as a single dose
- 18-24kg: 3 sprays as a single dose
- 24-30kg: 4 sprays as a single dose

1600microgram/actuation
- 30-40kg: 2 sprays as a single dose
- 40-50kg: 3 sprays as a single dose

Breakthrough
By buccal, subcutaneous or IV routes
- For breakthrough pain use 5-10% of total daily diamorphine dose every 1-4 hours as needed.

Dyspnoea
By buccal, subcutaneous or IV routes
- Dose as for pain, but at 50% of breakthrough dose

Notes:
- Diamorphine injection is licensed for the treatment of children who are terminally ill.
- For intranasal or buccal administration of diamorphine use the injection powder reconstituted in water for injections (unlicensed route of administration) or the nasal spray may be used (licensed for use in the management of severe acute pain from 2 years of age)
- In neonates, dosage interval should be extended to 6 or 8 hourly depending on renal function and the dose carefully checked, due to increased sensitivity to opioids in the first year of life.
- In poor renal function, dosage interval may be lengthened, or opioids only given as required and titrated against symptoms. Consider changing to fentanyl.
- For CSCI dilute with water for injections, as concentration incompatibility occurs with 0.9% saline at above 40 mg/ ml.
- Diamorphine can be given by subcutaneous infusion up to a strength of 250mg/ml
- Morphine injection is rapidly taking over from diamorphine, as the only benefit of diamorphine over morphine is its better solubility when high doses are needed and this is rarely a problem in paediatric doses.
- Available as : injection (5 mg, 10 mg, 30 mg, 100 mg, 500 mg ampoules); nasal spray 720microgram/actuation and 1600microgram/actuation (Ayendi Nasal Spray$^{(R)}$).

Evidence: [1, 2, 6, 33, 77, 78]

Diazepam

Use:
- Short term anxiety relief
- Agitation
- Panic attacks
- Relief of muscle spasm
- Treatment of status epilepticus.

Dose and routes

Short term anxiety relief, panic attacks and agitation
By mouth:
- **Child 2–12 years:** 1-2mg 3 times daily,
- **Child 12–18 years:** initial dose of 2 mg 3 times daily increasing as necessary and as tolerated to a maximum of 10 mg 3 times daily.

Relief of muscle spasm
By mouth:
- **Child 1–12 months:** initial dose of 250 microgram/kg twice a day,
- **Child 1–5 years:** initial dose of 2.5 mg twice a day,
- **Child 5–12 years:** initial dose of 5 mg twice a day,
- **Child 12–18 years:** initial dose of 10 mg twice a day; maximum total daily dose 40 mg.

Status epilepticus
By IV injection over 3–5 minutes:
- **Neonate:** 300-400 micrograms/kg as a single dose repeated once after 10 minutes if necessary
- **Child 1 month – 12 years:** 300-400 micrograms per kg (max 10 mg) repeated once after 10 minutes if necessary
- **Child 12–18 years:** 10 mg repeated once after 10 minutes if necessary (In hospital up to 20 mg as single dose may be used).

By rectum (rectal solution):
- **Neonate:** 1.25–2.5 mg repeated once after 10 minutes if necessary,
- **Child 1 month–2 years:** 5 mg repeated once after 10 minutes if necessary,
- **Child 2–12 years:** 5–10 mg repeated once after 10 minutes if necessary,
- **Child 12–18 years:** 10 mg repeated once after 10 minutes if necessary (in hospital up to 20 mg as a single dose may be used).

Notes
- Do not use in acute or severe respiratory insufficiency unless in the imminently dying

- Rectal tubes not licensed for children < 1 year old.
- Use with caution in mild-moderate hepatic disease
- Metabolised via the cytochrome P450 group of liver enzymes – potential for interaction with any concurrent medicine that induces or inhibits this group of enzymes. Enhancement of the central depressive effect may occur if diazepam is combined with drugs such as neuroleptics, antipsychotics, tranquillisers, antidepressants, hypnotics, analgesics, anaesthetics, barbiturates and sedative antihistamines
- Can cause dose-dependent drowsiness and impaired psychomotor and cognitive skills
- Almost 100% bioavailable when given orally or by rectal solution
- Onset of action ~15 minutes given orally and within 1-5 minutes given intravenously. Given as rectal solution, diazepam is rapidly absorbed from the rectal mucosa with maximum serum concentration reached within 17 minutes.
- Long plasma half-life of 24-48 hours with the active metabolite, nordiazepam, having a plasma half-life of 48-120 hours
- The oral solution may be administered via a gastrostomy tube. For administration via a jejunostomy tube, consider using tablets dispersed in water to reduce osmolarity.
- Available as: tablets (2 mg, 5 mg, 10 mg), oral solution (2 mg/5 mL, 5 mg/5 mL), rectal tubes (2.5 mg, 5 mg, 10 mg), and injection (5 mg/mL solution and 5 mg/ml emulsion).

Evidence: [1, 2, 6, 12, 19, 25, 52, 79-84]

Diclofenac Sodium

Use:
- Mild to moderate pain and inflammation, particularly musculoskeletal disorders.

Dose and routes
By mouth or rectum:
- **Child 6 months - 18 years:** initial dose of 0.3 mg/kg 3 times daily increasing if necessary to a maximum of 1 mg/kg 3 times daily (maximum 50 mg single dose).

By IM or IV infusion:
- **Child 2–18 years:** initial dose of 0.3 mg/kg 1–2 times daily; maximum of 150 mg/day and for a maximum of 2 days.

Notes:

Will cause closure of ductus arteriosus; contraindicated in duct dependent congenital heart disease
- Not licensed for use in children under 1 year; *suppositories* not licensed for use in children under 6 years except for use in children over 1 year for juvenile idiopathic arthritis; solid dose forms containing more than 25 mg not licensed for use in children; *injection (for IM bolus or IV infusion only)* not licensed for use in children.
- The risk of cardiovascular events secondary to NSAID use is undetermined in children. In adults, all NSAID use (including cyclo-oxygenase-2 selective inhibitors) can, to varying degrees, be associated with a small increased risk of thrombotic events (e.g. myocardial infarction and stroke) independent of baseline cardiovascular risk factors or duration of NSAID use; however, the greatest risk may be in those patients receiving high doses long term. A small increased thrombotic risk cannot be excluded in children.
- All NSAIDs are associated with gastro-intestinal toxicity. In adults, evidence on the relative safety of NSAIDs indicates differences in the risks of serious upper gastro-intestinal side-effects—piroxicam and ketorolac are associated with the highest risk; indometacin, diclofenac, and naproxen are associated with intermediate risk, and ibuprofen with the lowest risk (although high doses of ibuprofen have been associated with intermediate risk)

- Smallest dose that can be given practically by rectal route is 3.125 mg by cutting a 12.5 mg suppository into quarters (CC).
- For IV infusion, dilute in 5% glucose or 0.9% NaCl and infuse over 30-120 minutes
- Dispersible tablets may be administered via an enteral feeding tube. Disperse immediately before administration.
- Available as: tablets (25 mg, 50 mg, and 75 mg modified release), dispersible tablets (10 mg from a 'specials' manufacturer, 50 mg), modified release capsules (75mg and 100mg), injection (25 mg/mL Voltarol[R] for IM injection or IV infusion only), and suppositories (12.5 mg, 25 mg, 50 mg and 100 mg).

Evidence: [2, 6, 12, 43]

Dihydrocodeine

Use:
- Mild to moderate pain in patients known to be able to benefit.

Dose and routes
By mouth or deep subcutaneous or intramuscular injection:
- **Child 1-4 years:** 500microgram/kg every 4-6 hours,
- **Child 4-12 years:** initial dose of 500microgram/kg (maximum 30 mg/dose) every 4-6 hours. Dose may be increased if necessary to 1 mg/kg every 4-6 hours (maximum 30 mg/dose),
- **Child 12-18 years:** 30 mg (maximum 50 mg by intramuscular or deep subcutaneous injection) every 4-6 hours,
- Modified release tablets used 12 hourly (use ½ of previous total daily dose for each modified release dose).

Notes:
- Most preparations not licensed for children under 4 years.
- Relatively constipating compared with morphine / diamorphine and has a ceiling analgesic effect.
- Dihydrocodeine is itself an active substance, not a pro-drug like codeine.

- Oral bioavailability 20%, so probably equipotent with codeine by mouth (but opinion varies), twice as potent as codeine by injection.
- Time to onset 30 minutes, duration of action 4 hours for immediate release tablets.
- Side effects as for other opioids, plus paralytic ileus, abdominal pain, paraesthesia.
- Precautions: avoid or reduce dose in hepatic or renal failure.
- Available as: tablets (30 mg, 40 mg), oral solution (10 mg/5 mL), injection (CD) (50 mg/mL 1 mL ampoules) and m/r tablets (60 mg, 90 mg, 120 mg).

Evidence: [2, 5, 33, 43] EA, CC for injection

Docusate

Use:
- Constipation (faecal softener).

Dose and routes
By mouth:
- **Child 6 months–2 years:** initial dose of 12.5 mg 3 times daily; adjust dose according to response,
- **Child 2–12 years:** initial dose of 12.5 mg 3 times daily. Increase to 25 mg 3 times daily as necessary and then further adjust dose according to response,
- **Child 12–18 years:** initial dose 100mg/dose 3 times daily. Adjust as needed according to response up to 500 mg/day in divided doses.

By rectum:
- **Child 12–18 years:** 1 enema as single dose.

Notes:
- Adult oral solution and capsules not licensed in children < 12 years.

- Oral preparations act within 1–2 days.
- Rectal preparations act within 20 mins.
- Mechanism of action is emulsifying, wetting and mild stimulant.
- Doses may be exceeded on specialist advice.
- Available as capsules (100 mg), oral solution (12.5 mg/5 mL paediatric, 50 mg/5 mL adult), and enema (120 mg in 10 g single dose pack).

Evidence: [2]

Domperidone

MHRA April 2014: Domperidone is associated with a small increased risk of serious cardiac side effects. Its use is now restricted to the relief of symptoms of nausea and vomiting and the dosage and duration of use have been reduced. Domperidone is now **contraindicated** for use in those with underlying cardiac conditions and other risk factors.
The indications and doses below are therefore largely unlicensed usage in a particular population. Use the minimum effective dose. Do not use in those with known cardiac problems or other risk factors.

Use:
- Nausea and vomiting where poor GI motility is the cause.
- Gastro-oesophageal reflux resistant to other therapy.

Dose and routes
For nausea and vomiting
By mouth:
- **> 1 month and body-weight ≤ 35 kg:** initial dose of 250microgram/kg 3–4 times daily increasing if necessary to 500microgram/kg 3-4 times daily. Maximum 2.4 mg/kg (or 80mg) in 24 hours,
- **Body-weight > 35 kg:** initial dose of 10 mg 3-4 times daily increasing if necessary to 20 mg 3-4 times daily. Maximum 80 mg in 24 hours.

By rectum:
- **Body-weight 15–35 kg:** 30 mg twice a day,
- **Body-weight > 35 kg:** 60 mg twice a day.

For gastro-oesophageal reflux and gastrointestinal stasis
By mouth:
- **Neonate:** initial dose of 100microgram/kg 4–6 times daily before feeds. Dose may be increased, if necessary, to maximum of 300microgram/kg 4-6 times daily,
- **Child 1 month–12 years:** initial dose of 200microgram/kg (maximum single dose 10 mg) 3-4 times daily before food. Dose may be increased, if necessary, to 400microgram/kg 3-4 times daily. Maximum single dose 20 mg,
- **Child 12–18 years:** initial dose of 10 mg 3–4 times daily before food. Dose may be increased, if necessary, to 20 mg 3-4 times daily.

Notes

- Domperidone may be associated with an increased risk of serious ventricular arrhythmias or sudden cardiac death.
- Domperidone is contraindicated in those
  - With conditions where cardiac conduction is, or could be, impaired
  - With underlying cardiac diseases such as congestive heart failure
  - Receiving other medications known to prolong QT interval (e.g. erythromycin, ketoconazole) or which are potent CYP3A4 inhibitors
  - With severe hepatic impairment
- This risk may be higher with daily doses greater than 30mg. Use at lowest effective dose.
- Not licensed for use in gastro-intestinal stasis; not licensed for use in children for gastro-oesophageal reflux disease.
- Reduced ability to cross blood brain barrier, so less likely to cause extrapyramidal side effects compared with metoclopramide.

- Promotes gastrointestinal motility so diarrhoea can be an unwanted (or useful) side effect.
- Not to be used in patients with hepatic impairment.
- For administration via an enteral feeding tube: Use the suspension formulation, although the total daily dose of sorbitol should be considered. If administering into the jejunum, dilute the suspension with at least an equal volume of water immediately prior to administration.
- Available as: tablets (10 mg), oral suspension (5 mg/5 mL), and suppositories (30 mg).

Evidence: [2, 3, 6, 12, 85-90]

Entonox (nitrous oxide)

Use:
- As self-regulated analgesia without loss of consciousness.
- Particularly useful for painful dressing changes.

Dose and routes
By inhalation:
- **Child usually > 5 years old:** self-administration using a demand valve. Up to 50% in oxygen according to child's needs.

Notes:
- Is normally used as a light anaesthesic.
- Rapid onset and then offset.
- Should only be used as self-administration using a demand valve; all other situations require a specialist paediatric anaesthetist.
- Use is dangerous in the presence of pneumothorax or intracranial air after head injury.
- Prolonged use can cause megaloblastic anaemia.
- May be difficult to make available in hospice settings especially if needed infrequently, due to training, governance and supply implications.

Evidence: [2, 91]

Erythromycin

Use:
- Gastrointestinal stasis (motilin receptor agonist).

Dose and routes
By mouth:
- **Neonate:** 3 mg/kg 4 times daily,
- **Child 1 month–18 years:** 3 mg/kg 4 times daily.

Notes:
- Not licensed for use in children with gastrointestinal stasis.
- Erythromycin is excreted principally by the liver, so caution should be exercised in administering the antibiotic to patients with impaired hepatic function or concomitantly receiving potentially hepatotoxic agents.
- Erythromycin is a known inhibitor of the cytochrome P450 system and may increase the serum concentration of drugs which are metabolised by this system. Appropriate monitoring should be undertaken and dosage should be adjusted as necessary. Particular care should be taken with medications known to prolong the QT interval of the electrocardiogram.
- Available as: tablets (250 mg, 500 mg) and oral suspension (125 mg/5 mL, 250 mg/5 mL, 500mg/5 mL).

Evidence: [2, 92, 93] SR

Etamsylate

Use:
- Treatment of haemorrhage, including surface bleeding from ulcerating tumours.

Dose and routes
By mouth:
- **> 18 years:** 500 mg 4 times daily, indefinitely or until a week after cessation of bleeding.

Notes:
- Not licensed for use with children with haemorrhage.
- Available as: tablets (500 mg).

Evidence: [1]

Etoricoxib

Uses:
- Anti-inflammatory analgesic; adjuvant for musculoskeletal pain

Dose and route:
Oral:
- **Child 12-16 years:** initial dose of 30mg once daily. Dose may be increased as necessary and as tolerated to a maximum of 60mg once daily,
- **Child 16 years and older:** usual dose of 30-90mg once daily. Doses up to 120mg have been used on a short term basis in acute gouty arthritis in adults.

Notes:
- Oral selective cyclo-oxygenase (COX-2) inhibitor.
- Etoricoxib is not licensed for use in children less than 16 years of age. The pharmacokinetics of etoricoxib in children less than 12 years of age have not been studied.
- Etoricoxib may mask fever and other signs of inflammation.
- All NSAIDs should be used with caution in children with a history of hypersensitivity to any NSAID or in those with a coagulation disorder.
- Etoricoxib is contraindicated in those with: active peptic ulceration or active GI bleeding; severe hepatic or renal dysfunction; inflammatory bowel disease or congestive heart failure
- The risk of cardiovascular events secondary to NSAID use is undetermined in children. In adults COX-2 selective inhibitors, diclofenac (150mg daily) and ibuprofen (2.4g daily) are associated

- with an increased risk of thrombotic effects (e.g. myocardial infarction and stroke).
- All NSAIDs are associated with GI toxicity. In adults evidence on the relative safety of NSAIDs indicates differences in the risks of serious upper GI side-effects with piroxicam and ketorolac associated with the highest risk and ibuprofen at low to medium dose with the lowest risk. Selective COX-2 inhibitors are associated with a lower risk of serious upper GI side-effects than non-selective NSAIDs. Children appear to tolerate NSAIDs better than adults and GI side-effects are less common although they do still occur.
- Common (1-10% patients) AEs: alveolar osteitis; oedema/fluid retention; dizziness, headache; palpitations, arrhythmia; hypertension; bronchospasm; abdominal pain; constipation, flatulence, gastritis, heartburn/acid reflux, diarrhoea, dyspepsia/epigastric discomfort, nausea, vomiting, oesophagitis, oral ulcer; ALT increased, AST increased; ecchymosis; asthenia/fatigue, flu-like disease
- Potential drug interactions include warfarin (increase in INR); diuretics, ACE inhibitors and angiotensin II antagonists (increased risk of compromised renal function). Etoricoxib does NOT appear to inhibit or induce CYP isoenzymes. However, the main pathway of etoricoxib metabolism is dependent on CYP enzymes (primarily CYP3A4) so co-administration with drugs that are inducers or inhibitors of this pathway may affect the metabolism of etoricoxib.
- Etoricoxib is administered orally and may be taken with or without food. However, the onset of effect may be faster when administered without food.
- Etoricoxib tablets may be dispersed in 10ml water and will disintegrate to give fine granules that settle quickly but disperse easily and flush down an 8Fr NG or gastrostomy tube without blockage
- Available as: film coated tablets 30mg, 60mg, 90mg, 120mg. Tablets contain lactose

Evidence: [1, 94] SR EA

Fentanyl

Use:
- Step 2 WHO pain ladder once dose is titrated.

Dose and routes
Normally convert using oral morphine equivalent (OME) from previous analgesia.
Use the following **starting** doses in the opioid naive patient. The maximum dose stated applies to **starting** dose only.

By transmucosal application (lozenge with oromucosal applicator),
- **Child 2-18 years and greater than 10 kg:** 15 micrograms/kg as a single dose, titrated to a maximum dose 400 micrograms (higher doses under specialist supervision).

By intranasal
- **Neonate - Child<2 years:** 1 microgram/kg as a single dose,
- **Child 2-18 years:** 1-2 micrograms/kg as a single dose, with initial maximum single dose of 50 micrograms.

By transdermal patch or continuous infusion:
- Based on oral morphine dose equivalent (given as 24-hour totals).

By intravenous injection
- **Neonate or infant:** 1-2 micrograms/kg per dose slowly over 3-5 minutes; repeated every 2-4 hours,
- **Child:** 1-2 micrograms/kg per dose, repeated every 30-60 minutes.

By continuous intravenous infusion
- **Neonate or infant:** initial IV bolus of 1-2 micrograms/kg (slowly over 3-5 minutes) followed by 0.5-1 microgram/kg/hour,
- **Child:** initial IV bolus of 1-2 micrograms/kg (slowly over 3-5 minutes) followed by 0.5-1 microgram/kg/hour.

72-hour Fentanyl patches are *approximately* equivalent to the following 24-hour doses of oral morphine

morphine salt 30 mg daily ≡ fentanyl '12' patch
morphine salt 60 mg daily ≡ fentanyl '25' patch
morphine salt 120 mg daily ≡ fentanyl '50' patch
morphine salt 180 mg daily ≡ fentanyl '75' patch
morphine salt 240 mg daily ≡ fentanyl '100' patch

Notes:
- Fentanyl patch should be changed every 72 hours and the site of application rotated.
- Injection not licensed for use in children less than 2 years of age. Lozenges and nasal sprays are not licensed for use in children.
- The injection solution can be administered by the intranasal route for doses less than 50 micrograms which is the lowest strength of nasal spray available.
- Injection solution could be administered drop wise (may be unpleasant) or using an atomiser device that A+E units use for intranasal diamorphine.
- The main advantage of fentanyl over morphine in children is its availability as a transdermal formulation.
- It can simplify analgesic management in patients with poor, deteriorating or even absent renal function.
- Avoid or reduce dose in hepatic impairment.
- It is a synthetic opioid, very different in structure from morphine, and therefore ideal for opioid switching.
- Evidence that it is less constipating than morphine has not been confirmed in more recent studies [95].
- The patch formulation is not usually suitable for the initiation or titration phases of opioid management in palliative care since the patches represent large increments and because of the time lag to achieve steady state.
- The usefulness of lozenges in children is limited by the dose availability and no reliable conversion factor which also varies between preparations. Another caution is that opioid morphine approximate equivalence of the smallest lozenge (200 microgram)

is 30 mg, meaning it is probably suitable to treat breakthrough pain only for children receiving a total daily dose equivalent of 180 mg morphine or more. Older children will often choose to remove the lozenge before it is completely dissolved, giving them some much-valued control over their analgesia. Note: the lozenge must be rotated in buccal pouch, not sucked. Unsuitable for pain in advanced neuromuscular disorders where independent physical rotation of lozenge is not possible.
- Pharmacokinetics of fentanyl intranasally are favourable but it is not always practical and/or well tolerated in children.

Available as fentanyl citrate:
- Intranasal spray (50 micrograms/metered spray, 100 micrograms/metered spray, 200 micrograms/metered spray Instanyl[R]). Also available as PecFent 100 microgram/metered spray and 400 microgram/metered spray.
- Lozenge with oromucosal applicator (200 micrograms, 400 micrograms, 600 micrograms, 800 micrograms, 1.2 mg, 1.6 mg Actiq[R]).
- Sublingual tablets (100, 200, 300, 400, 600 and 800micrograms (Abstral[(R)])) and buccal tablets (Effentora[(R)] 100, 200, 400, 600 and 800micrograms; Breakyl[(R)] 200, 400, 600, 800 and 1200micrograms)
- Patches (12 microgram/hour, 25 microgram/hour, 50 microgram/hour, 75 microgram/hour, 100 microgram/hour).

Evidence: [2, 4, 5, 8, 77, 96-107] CC

Fluconazole

Use:
- Mucosal candidiasis infection, invasive candidal infections or prevention of fungal infections in immunocompromised patients.

Dose and routes

Mucosal candidal infection
By mouth or intravenous infusion:
- **Neonate under 2 weeks:** 3-6 mg/kg on first day then 3 mg/kg every 72 hours,
- **Neonate between 2-4weeks:** 3-6 mg/kg on first day then 3 mg/kg every 48 hours,
- **Child 1 month–12 years:** 3-6 mg/kg on first day then 3 mg/kg (maximum 100 mg) daily,
- **Child 12–18 years:** 50 mg/day. Increase to 100 mg/day in difficult infections.

Invasive candidal infections and cryptococcal infections
By mouth or intravenous infusion:
- **Neonate under 2 weeks:** 6-12mg/kg every 72 hours
- **Neonate over 2 weeks:** 6-12mg/kg every 48 hours
- **Child 1 month – 18 years:** 6-12mg/kg (max.800mg) every 24 hours

Prevention of fungal infections in immunocompromised patients
By mouth or intravenous infusion
- **Neonate under 2 weeks:** 3-12mg/kg every 72 hours
- **Neonate over 2 weeks:** 3-12mg/kg every 48 hours
- **Child 1 month – 18 years:** 3-12mg/kg (max.400mg) every 24 hours

Notes:
- Use for 7-14 days in oropharyngeal candidiasis.
- Use for 14-30 days in other mucosal infection.
- Different duration of use in severely immunocompromised patients.
- Fluconazole is a potent CYP2C9 inhibitor and a moderate CYP3A4 inhibitor. Fluconazole is also an inhibitor of CYP2C19. Fluconazole treated patients who are concomitantly treated with medicinal products with a narrow therapeutic window metabolised through CYP2C9, CYP2C19 and CYP3A4, should be monitored.

- The most frequently (>1/10) reported adverse reactions are headache, abdominal pain, diarrhoea, nausea, vomiting, alanine aminotransferase increased, aspartate aminotransferase increased, blood alkaline phosphatase increased and rash.
- For *intravenous infusion*, give over 10–30 minutes; do not exceed an infusion rate of 5–10 mL/minute
- Oral suspension may be administered via an enteral feeding tube.
- Available as: capsules (50 mg, 150 mg, 200 mg); oral suspension (50 mg/5 mL, 200 mg/mL) and IV infusion (2mg/ml in 2ml, 50ml or 100ml infusion bags).

Evidence: [2, 12, 108, 109]

Fluoxetine

Use:
- Major depression.

Dose and routes
By mouth:
- **Child 8–18 years:** initial dose 10 mg once a day. May increase after 1-2 weeks if necessary to a maximum of 20 mg once daily.

Notes:
- Licensed for use in children from 8 years of age.
- Use with caution in children ideally with specialist psychiatric advice.
- Increased risk of anxiety for first 2 weeks.
- Onset of benefit 3-4 weeks.
- Consider long half-life when adjusting dosage. Do not discontinue abruptly.
- May also help for neuropathic pain and intractable cough.
- Suicide related behaviours have been more frequently observed in clinical trials among children and adolescents treated with antidepressants compared with placebo. Mania and hypomania have been commonly reported in paediatric trials.

- The most commonly reported adverse reactions in patients treated with fluoxetine were headache, nausea, insomnia, fatigue and diarrhoea. Undesirable effects may decrease in intensity and frequency with continued treatment and do not generally lead to cessation of therapy.
- Because the metabolism of fluoxetine, (like tricyclic antidepressants and other selective serotonin antidepressants), involves the hepatic cytochrome CYP2D6 isoenzyme system, concomitant therapy with drugs also metabolised by this enzyme system may lead to drug interactions.
- Must not be used in combination with a MAOI.
- Available as: capsules (20 mg) and oral liquid (20 mg/5 mL).

Evidence: [1, 2, 110-117]

Gabapentin

Use:
- Adjuvant in neuropathic pain.

Dose and routes
By mouth:
- **Child >2years:**
    - Day 1 10 mg/kg as a single dose (maximum single dose 300 mg),
    - Day 2 10 mg/kg twice daily (maximum single dose 300 mg),
    - Day 3 onwards 10 mg/kg three times daily (maximum single dose 300 mg),
    - Increase further if necessary to maximum of 20 mg/kg/dose (maximum single dose 600 mg).
- **From 12 years:** the maximum daily dose can be increased according to response to a maximum of 3600 mg/day.

Notes:
- Not licensed for use in children with neuropathic pain.

- Speed of titration after first 3 days varies between increases every 3 days for fast regime to increase every one to two weeks in debilitated children or when on other CNS depressants.
- No consensus on dose for neuropathic pain. Doses given based on doses for partial seizures and authors' experience.
- Dose reduction required in renal impairment. Consult manufacturer's literature.
- Very common (>1 in 10) side-effects: somnolence, dizziness, ataxia, viral infection, fatigue, fever
- Capsules can be opened but have a bitter taste.
- Available as: capsules (100 mg, 300 mg, 400 mg); tablets (600 mg, 800 mg) and oral solution 50mg in 5ml (Rosemont – however this product contains propylene glycol, acesulfame K and saccharin sodium and levels may exceed the recommended WHO daily intake limits if high doses are given to adolescents with low body-weight (39–50 kg)).

Evidence: [1, 2, 31, 33, 118-120] CC, SR

Gaviscon®

Use:
- Gastro-oesophageal reflux, dyspepsia, and heartburn.

Dose and routes
By mouth:
- **Neonate–2 years, body weight < 4.5 kg:** 1 dose (half dual sachet) when required mixed with feeds or with water for breast fed babies, maximum 6 doses in 24 hours,
- **Neonate–2 years body weight > 4.5 kg:** 2 doses (1 dual sachet) when required mixed with feeds or with water for breast fed babies or older infants, maximum 6 doses in 24 hours,

Gaviscon Liquid
- **Child 2-12 years:** 1 tablet or 5-10ml liquid after meals and at bedtime
- **Child 12-18 years:** 1-2 tablets or 10-20ml after meals and at bedtime

Gaviscon Advance
- **Child 2-12 years:** 1 tablet or 2.5-5ml after meals and at bedtime (under medical advice only)
- **Child 12-18 years:** 1-2 tablets or 5-10ml suspension after meals and at bedtime

Notes:
- Gaviscon Infant Sachets licensed for infants and young children up to 2 years of age but use <1 year only under medical supervision. Gaviscon liquid and tablets – licensed for use from 2 years of age but age 2-6 years only on medical advice. Gaviscon Advance suspension and tablets licensed for use from 12 years of age; under 12 years on medical advice only.
- Gaviscon Infant should not to be used with feed thickeners, nor with excessive fluid losses, (e.g. fever, diarrhoea, vomiting).
- Gaviscon Liquid contains 3.1mmol sodium per 5ml; Gaviscon tablets contain 2.65mmol sodium and also contain aspartame. Gaviscon Advance Suspension contains 2.3mmol sodium and 1mmol potassium per 5ml and 2.25mmol sodium and 1mmol potassium per 5ml and also contain aspartame. Gaviscon Infant Sachets contain 0.92mmol sodium per dose (half dual sachet).
- Available as: Gaviscon liquid and tablets; Gaviscon Advance suspension and tablets; and infant sachets (comes as dual sachets, each half of dual sachet is considered one dose).

Evidence: [1-3]

Glycerol (glycerin)

Use:
- Constipation.

Dose and routes
By rectum:
- **Neonate:** tip of a glycerol suppository (slice a small chip of a 1 g suppository with a blade),
- **Child 1 month–1 year:** 1 g infant suppository as required,
- **Child 1–12 years:** 2 g child suppository as required,
- **Child 12–18 years:** 4 g adult suppository as required.

Notes:
- Moisten with water before insertion.
- Hygroscopic and lubricant actions. May also be a rectal stimulant.
- Response usually in 20 minutes to 3 hours.
- Available as: suppositories (1 g, 2 g, and 4 g).

Evidence: [1, 2, 43] CC

Glycopyrronium bromide

Use:
- Control of upper airways secretion and hypersalivation.

Dose and routes
By mouth:
- **Child 1 month-18 years:** initial dose of 40 microgram/kg 3–4 times daily. The dose may be increased as necessary to 100 microgram/kg 3-4 times daily.
  Maximum 2 mg/dose given 3 times daily.

Subcutaneous:
- **Child 1 month-12 years:** initial dose of 4 micrograms/kg 3 to 4 times daily. The dose may be increased as necessary to 10 microgram/kg 3-4 times daily.
  Maximum 200 microgram/dose given 4 times daily,
- **Child 12-18 years:** 200 micrograms every 4 hours when required.

Continuous subcutaneous infusion:
- **Child 1 month -12 years:** initial dose of 10 micrograms/kg/24 hours. The dose may be increased as necessary to 40 microgram/kg/24 hours
  (maximum 1.2 mg/24 hours),
- **Child 12-18 years:** initial dose of 600micrograms /24 hours. The dose may be increased as necessary to 1.2 mg/24 hours. Maximum recommended dose is 2.4 mg/24 hours.

Notes:
- Not licensed for use in children for control of upper airways secretion and hypersalivation.
- Excessive secretions can cause distress to the child, but more often cause distress to those around him.
- Treatment is more effective if started before secretions become too much of a problem.
- Glycopyrronium does not cross the blood brain barrier and therefore has fewer side effects than hyoscine hydrobromide, which is also used for this purpose. Also fewer cardiac side effects.
- Slower onset response than with hyoscine hydrobromide or butylbromide.
- Oral absorption of glycopyrronium is very poor with wide inter-individual variation.
- For oral administration injection solution may be given or the tablets may be crushed and suspended in water. For administration via an enteral feeding tube, tablets may be dispersed in water. However the coarse dispersion settles quickly. Flush syringe and tube well to ensure all the dose is received.

- Administration by CSCI: good compatibility data available with other commonly used palliative agents.
- Available as: tablets (1 mg, 2 mg via an importation company as the tablets are not licensed in the UK): dosing often too inflexible for children, costly and can be difficult to obtain. Injection (200 microgram/mL 1mL ampoules) can also be used orally (unlicensed route).Oral solution can also be prepared extemporaneously from glycopyrronium powder and obtained from a 'specials' manufacturer.

Evidence: [2, 121-123]

Haloperidol

Use:
- Nausea and vomiting where cause is metabolic or in difficult to manage cases such as end stage renal failure.
- Restlessness and confusion.
- Intractable hiccups.
- Psychosis, hallucination

Dose and routes
By mouth for *nausea and vomiting*:
- **Child 1 month–12 years:** initial dose of 50 microgram/kg/24 hours (initial maximum 3mg/24hrs) in divided doses. The dose may be increased as necessary to a maximum of 170 microgram/kg/24 hours in divided doses,
- **Child 12–18 years:** 1.5 mg once daily at night, increasing as necessary to 1.5 mg twice a day; maximum 5 mg twice a day.

By mouth for *restlessness and confusion*:
- **Child 1 month–12 years:** initial dose of 50 microgram/kg/24 hours (initial maximum 3mg/24hrs) in divided doses. The dose

may be increased as necessary to a maximum of 170 microgram/kg/24 hours in divided doses
- **Child 12-18 years:** 10–20 microgram/kg every 8–12 hours; maximum 10 mg/day.

By mouth for *intractable hiccups:*
- **Child 1 month–12 years:** initial dose of 50 microgram/kg/24 hours (initial maximum 3mg/24hrs) in divided doses. The dose may be increased as necessary to a maximum of 170 microgram/kg/24 hours in divided doses
- **Child 12–18 years:** 1.5 mg 3 times daily.

By continuous IV or SC infusion (for any indication):
- **Child 1 month–12 years:** initial dose of 25 microgram/kg/24 hours (initial maximum 1.5mg/24hrs). The dose may be increased as necessary to a maximum of 85 microgram/kg/24 hours,
- **Child 12–18 years:** initial dose of 1.5 mg/24 hours. The dose may be increased as necessary to a suggested maximum of 5 mg/24 hours although higher doses may be used under specialist advice.

Notes:
- D2 receptor antagonist and typical antipsychotic.
- Not licensed for use in children with nausea and vomiting, restlessness and confusion or intractable hiccups. Injection is licensed only for IM administration in adults; IV and SC administration off-label (all ages).
- Haloperidol can cause potentially fatal prolongation of the QT interval and torsades de pointes particularly if given IV (off-label route) or at higher than recommended doses. Caution is required if any formulation of haloperidol is given to patients with an underlying predisposition e.g. those with cardiac abnormalities, hypothyroidism, familial long QT syndrome, electrolyte imbalance or taking other drugs known to prolong the QT interval. If IV haloperidol is essential, ECG monitoring during drug administration is recommended.
- Dosages for restlessness and confusion are often higher.

- Adult dosages can exceed 15 mg/24 hours in severe agitation.
- Oral doses are based on an oral bioavailability of ~50% of the parenteral route i.e. oral doses ~2x parenteral.
- Useful as long acting – once daily dosing is often adequate.
- Oral solutions may be administered via an enteral feeding tube.
- Available as: tablets (500 microgram, 1.5 mg, 5 mg, 10 mg, 20 mg), capsules (500 microgram), oral liquid (1 mg/mL, 2 mg/mL), and injection (5 mg/mL).

Evidence: [1, 2, 6, 12, 75, 124-128]

Hydromorphone

Use:
- Alternative opioid analgesic for severe pain especially if intolerant to other strong opioids.
- Antitussive.

Dose and routes
Normally convert using oral morphine equivalent (OME) from previous analgesia.
Use the following **starting** doses in opioid naive patient. The maximum dose stated applies to **starting** dose only.

By mouth:
- **Child 1–18 years:** 30 micrograms/ kg per dose maximum 2mg per dose every 3-4 hours increasing as required. Modified release capsules with an initial dose of 4mg every 12 hours may be used from 12 years of age.

By IV or SC injection:
- **Child 1-18 years**: initially 15 micrograms/kg per dose slowly over at least 2-3 minutes every 3-6 hours.
- Convert from oral (halve dose for equivalence).

Notes:
- Hydrated morphine ketone effects are common to the class of mu agonist analgesics.
- Injection is not licensed in the UK. May be possible to obtain via a specialist importation company but as hydromorphone is a CD this is not a straightforward process.
- Oral form licensed for use in children from 12 years of age with cancer pain.
- Oral bioavailability 37-62% (wide inter-individual variation), onset of action 15 min for SC, 30 min for oral. Peak plasma concentration 1 hour orally. Plasma half life 2.5 hours early phase, with a prolonged late phase. Duration of action 4-5 hours.
- Potency ratios seem to vary more than for other opioids. This may be due to inter-individual variation in metabolism or bioavailability.
- Conversion of oral morphine to oral hydromorphone: divide morphine dose by 5-7
- Conversion of IV Morphine to IV hydromorphone: Divide morphine dose by 5-7
- Dosage discontinuation: after short-term therapy (7–14 days), the original dose can be decreased by 10–20% of the original dose every 8 hours increasing gradually the time interval. After long-termtherapy, the dose should be reduced not more than 10–20% per week.
- Caution in hepatic impairment, use at reduced starting doses
- Modified release capsules are given 12 hourly.
- Capsules (both types) can be opened and contents sprinkled on soft food. Capsule contents must not however be administered via an enteral feeding tube as likely to cause blockage.
- Available as: capsules (1.3 mg, 2.6 mg) and modified release capsules (2 mg, 4 mg, 8 mg, 16 mg, 24 mg).

Evidence: CC, EA, [1, 2, 4, 5, 29, 33, 100, 101, 129, 130]

Hyoscine butylbromide

Use:
- Adjuvant where pain is caused by spasm of the gastrointestinal or genitourinary tract.
- Management of secretions, especially where drug crossing the blood brain barrier is an issue.

Dose and routes
By mouth or IM or IV injection:
- **Child 1 month-4 years:** 300–500 micrograms/kg (maximum 5 mg/dose) 3–4 times daily,
- **Child 5-12 years:** 5-10 mg 3–4 times daily,
- **Child 12-18 years:** 10–20 mg 3–4 times daily.

By continuous subcutaneous infusion
- **Child 1month- 4 years:** 1.5 mg/kg/24 hours (max 15 mg/24 hours),
- **Child 5-12 years:** 30 mg/24 hours,
- **Child 12-18 years:** up to 60-80mg/24 hours,
- Higher doses may be needed; doses used in adults range from 20-120 mg/24 hours (maximum dose 300 mg/24 hours).

Notes:
- Does not cross blood brain barrier (unlike hyoscine hydrobromide), hence no central antiemetic effect and doesn't cause drowsiness.
- Tablets are not licensed for use in children < 6 years old.
- Injection is not licensed for use in children.
- The injection solution may be given orally or via an enteral feeding tube. If the tube exits in the jejunum, consider using parenteral therapy. Injection solution can be stored for 24 hours in the refrigerator.
- IV injection should be given slowly over 1 minute and can be diluted with glucose 5% or sodium chloride 0.9%.
- Available as: tablets (10 mg) and injection (20 mg/mL).

Evidence: [1, 2, 12, 121, 123]

Hyoscine hydrobromide

Use:
- Control of upper airways secretions and hypersalivation.

Dose and routes
By mouth or sublingual:
- **Child 2–12 years:** 10 micrograms/kg (maximum 300 micrograms single dose) 4 times daily,
- **Child 12–18 years:** 300 micrograms 4 times daily.

By transdermal route:
- **Child 1 month–3 years:** quarter of a patch every 72 hours,
- **Child 3–10 years:** half of a patch every 72 hours,
- **Child 10–18 years:** one patch every 72 hours.

By SC or IV injection or infusion:
- **Child 1 month–18 years:** 10 micrograms/kg (maximum 600 micrograms) every 4–8 hours or CSCI/IV infusion 40-60microgram/kg/24 hours. Maximum suggested dose is 2.4 mg in 24 hours although higher doses are often used by specialist units.

Notes:
- Not licensed for use in children for control of upper airways secretion and hypersalivation.
- Higher doses often used under specialist advice.
- Can cause delirium or sedation (sometimes paradoxical stimulation) with repeated dosing. Constipating.
- Apply patch to hairless area of skin behind ear.
- The patch can cause alteration of the pupil size on the side it is placed.
- Some specialists advise that transdermal patches should not be cut – however, the manufacturers of Scopoderm TTS patch have confirmed that it is safe to do this.

- Injection solution may be administered orally.
- Available as: tablets (150 micrograms, 300 micrograms), patches (releasing 1 mg/72hours), and injection (400 microgram/mL, 600 microgram/mL). An oral solution is available via a 'specials' manufacturer.

Evidence: [1, 2, 43, 121-123]

Ibuprofen

Use:
- Simple analgesic
- Pyrexia
- Adjuvant for musculoskeletal pain.

Dose and routes
By mouth:
- **Neonate:** 5 mg/kg/dose every 12 hours
- **Child 1–3 months:** 5 mg/kg 3–4 times daily preferably after food,
- **Child 3–6 months:** 50 mg 3 times daily preferably after food; in severe conditions up to 30 mg/kg daily in 3–4 divided doses,
- **Child 6 months–1 year:** 50 mg 3–4 times daily preferably after food; in severe conditions up to 30 mg/kg daily in 3–4 divided doses,
- **Child 1-4 years:** 100 mg 3 times daily preferably after food. In severe conditions up to 30 mg/kg daily in 3–4 divided doses,
- **Child 4–7 years:** 150 mg 3 times daily, preferably after food. In severe conditions, up to 30 mg/kg daily in 3–4 divided doses.
- **Child 7–10 years:** 200 mg 3 times daily, preferably after food. In severe conditions, up to 30 mg/kg daily in 3–4 divided doses. Maximum daily dose 2.4 g,

- **Child 10–12 years:** 300 mg 3 times daily, preferably after food. In severe conditions, up to 30 mg/kg daily in 3–4 divided doses. Maximum daily dose 2.4 g,
- **Child 12-18 years:** 300-400 mg 3-4 times daily preferably after food. In severe conditions the dose may be increased to a maximum of 2.4 g/day.

Pain and inflammation in rheumatic diseases, including idiopathic juvenile arthritis:
- **Child aged 3 months–8 years and body weight > 5 kg:** 30–40 mg/kg daily in 3–4 divided doses preferably after food. Maximum 2.4 g daily.

In systemic juvenile idiopathic arthritis:
- Up to 60 mg/kg daily in 4–6 divided doses up to a maximum of 2.4 g daily (off-label).

Notes:
- **Will cause closure of ductus arteriosus; contraindicated in duct dependent congenital heart disease.**
- Orphan drug licence for closure of ductus arteriosus in preterm neonate.
- Not licensed for use in children under 3 months of age or weight less than 5kg.
- Topical preparations and granules are not licensed for use in children.
- **Ibuprofen** combines anti-inflammatory, analgesic, and antipyretic properties. It has fewer side-effects than other NSAIDs but its anti-inflammatory properties are weaker.
- The risk of cardiovascular events secondary to NSAID use is undetermined in children. In adults, all NSAID use (including cyclo-oxygenase-2 selective inhibitors) can, to varying degrees, be associated with a small increased risk of thrombotic events (e.g. myocardial infarction and stroke) independent of baseline cardiovascular risk factors or duration of NSAID use; however, the greatest risk may be in those patients receiving high doses

long term. A small increased thrombotic risk cannot be excluded in children.
- All NSAIDs are associated with gastro-intestinal toxicity. In adults, evidence on the relative safety of NSAIDs indicates differences in the risks of serious upper gastro-intestinal side-effects—piroxicam and ketorolac are associated with the highest risk; indometacin, diclofenac, and naproxen are associated with intermediate risk, and ibuprofen with the lowest risk (although high doses of ibuprofen have been associated with intermediate risk)
- Caution in asthma and look out for symptoms and signs of gastritis.
- Consider use of a proton pump inhibitor with prolonged use of ibuprofen.
- For administration via an enteral feeding tube, use a liquid preparation; dilute with an equal volume of water immediately prior to administration where possible.
- Available as: tablets (200 mg, 400 mg, 600 mg), capsule (300 mg MR), oral syrup (100 mg/5 mL), granules (600 mg/sachet), and spray, creams and gels (5%).

Evidence: [1-3, 12, 131],

Ipratropium Bromide

Use:
- Wheezing/ Breathlessness caused by bronchospasm

Dose and routes
Nebulised solution
- **Child less than 1 year:** 62.5 micrograms 3 to 4 times daily,
- **Child 1-5 years:** 125-250 micrograms 3 to 4 times daily,
- **Child 5-12 years:** 250-500 micrograms 3 to 4 times daily,
- **Child over 12 years:** 500 micrograms 3 to 4 times daily.

Aerosol Inhalation
- **Child 1month-6 years:** 20 micrograms 3 times daily,
- **Child 6-12 years:** 20-40 micrograms 3 times daily,
- **Child 12-18 years:** 20-40 micrograms 3-4 times daily.

Notes
- Inhaled product should be used with a suitable spacer device, and the child/ carer should be given appropriate training.
- In acute asthma, use via an oxygen driven nebuliser.
- In severe acute asthma, dose can be repeated every 20-30 minutes in first two hours, then every 4-6 hours as necessary.
- Available as: nebuliser solution (250 micrograms in 1mL, 500 micrograms in 2mL), aerosol inhaler (20 microgram per metered dose).

Evidence: RE [2, 6]

Ketamine

Use:
- Adjuvant to a strong opioid for neuropathic pain.
- To reduce N-methyl-D-aspartate (NMDA) receptor wind-up pain and opioid tolerance

Dose and routes
By mouth or sublingual:
- **Child 1 month – 12 years:** Starting dose 150 microgram/kg, as required or regularly 6 – 8 hourly: increase in increments of 150 microgram/kg up to 400 microgram/kg as required. Doses equivalent to 3 mg/kg have been reported in adults,
- **Over 12 years and adult:** 10 mg as required or regularly 6 – 8 hourly; increase in steps of 10 mg up to 50 mg as required. Doses up to 200 mg 4 times daily reported in adults.

By continuous SC or IV infusion:
- **Child 1 month – adult:** Starting dose 40 microgram/kg/hour. Increase according to response; usual maximum 100 microgram/kg/hour. Doses up to 1.5 mg/kg/hour in children and 2.5 mg/kg/hour in adults have been reported.

Notes:
- NMDA antagonist.
- Specialist use only.
- Not licensed for use in children with neuropathic pain.
- Higher doses (bolus injection 1–2 mg/kg, infusions 600 – 2700 microgram/kg/hour) used as an anaesthetic e.g. for short procedures.
- Sublingual doses should be prepared in a maximum volume of 2ml. The bitter taste may make this route unpalatable. Special preparations for sublingual use are available in UK.
- Enteral dose equivalents may be as low as 1/3 IV or SC dose because ketamine is potentiated by hepatic first pass metabolism. Other papers quote a 1:1 SC to oral conversion ratio (Ref: Benitez-Rosario)
- Agitation, hallucinations, anxiety, dysphoria and sleep disturbance are recognised side effects. These may be less common in children and when sub-anaesthetic doses are used.
- Ketamine can cause urinary tract symptoms- frequency, urgency, dysuria and haematuria. Consider discontinuing Ketamine if these symptoms occur. (new ref: Shahani, 2007; Chen)
- Caution in severe hepatic impairment, consider dose reduction.
- Dilute in 0.9% saline for subcutaneous or intravenous infusion
- Can be administered as a separate infusion or by adding to opioid infusion/ PCA/NCA.
- Can also be used intranasally and as a topical gel.
- Available as: injection (10 mg/mL, 50 mg/mL, 100 mg/mL) and oral solution 50 mg in 5 ml (from a 'specials' manufacturer). Injection solution may be given orally. Mix with a flavoured soft drink to mask the bitter taste.

Evidence: [101, 132-142] CC, EA

Ketorolac

Use:
- Short-term management of moderate-severe acute postoperative; limited evidence of extended use in chronic pain

Doses and routes:

Short-term management of moderate to severe acute postoperative pain (NB Licensed duration is a maximum of 2 days; not licensed for use in adolescents and children less than 16 years of age)

IV bolus (over at least 15 seconds) or IM bolus:
- **Child 1-16 years:** initially 0.5–1 mg/kg (max. 10 mg), then 500 micrograms/kg (max. 15 mg) every 6 hours as required; max. 60mg daily,
- **Child >16 years:** initially 10 mg, then 10–30 mg every 4–6 hours as required (up to every 2 hours during initial postoperative period); max. 90 mg daily (those weighing less than 50 kg max. 60 mg daily).

Chronic pain in palliative care (unlicensed indication; data limited and of poor quality. Anecdotal reports of effectiveness for patients with bone pain unresponsive to oral NSAIDs)

SC bolus
- **Child >16 years:** 15-30mg/dose three times daily

CSCI
- **Child >16 years:** initial dose of 60mg/24hours. Increase if necessary by 15mg/24hours to a maximum of 90mg/24hours

Notes:
- NSAID which has potent analgesic effects with only moderate anti-inflammatory action.

- Licensed only for the short-term management (maximum of 2 days) of moderate to severe acute postoperative pain in adults and adolescents from 16 years of age.
- SC administration is an unlicensed route of administration.
- Contraindications: previous hypersensitivity to ketorolac or other NSAIDs; history of asthma; active peptic ulcer or history of GI bleeding; severe heart, hepatic or renal failure; suspected or confirmed cerebrovascular bleeding or coagulation disorders. Do not use in combination with any other NSAID.
- Dose in adults with mild renal impairment should not exceed 60mg/day.
- All NSAIDs are associated with GI toxicity. In adults, evidence on the relative safety of NSAIDs indicates ketorolac and piroxicam are associated with the highest risk. Use the lowest effective dose for the shortest time. In addition, consider use in combination with a gastroprotective drug especially if ketorolac is used for a prolonged period (outside the licensed indication). Use of ketorolac in adults carries a 15 times increased risk of upper gastrointestinal complications, and a 3 times increased risk compared with other nonselective NSAIDs.
- In adults all NSAID use can, to varying degrees, be associated with a small increased risk of thrombotic effects. The risk of cardiovascular effects secondary to NSAID use is undetermined in children, but in adults, ketorolac is associated with the highest myocardial infarction risk of all NSAIDs.
- Other potential adverse effects; Very common (>10% patients): headache, dyspepsia, nausea, abdominal pain; Common (1-10% patients): dizziness, tinnitus, oedema, hypertension, anaemia, stomatitis, abnormal renal function, pruritus, purpura, rash, bleeding and pain at injection site. Risk of adverse effects likely to increase with prolonged use.
- Drug interactions include: anticoagulants (contraindicated as the combination may cause an enhanced anticoagulant effect); corticosteroids (increased risk of GI ulceration of bleeding); diuretics (risk of reduced diuretic effect and increase the risk of nephrotoxicity of NSAIDs); other potential nephrotoxic drugs.

- Onset of action IV/IM injection: 10-30mins; maximal analgesia achieved within 1-2 hours and median duration of effect 4-6 hours.
- SC injection can be irritant therefore dilute to the largest volume possible (0.9% NaCl suggested). Alkaline in solution so high risk of incompatibility mixed with acidic drugs. Some data of compatibility in 0.9% saline with diamorphine or oxycodone. Incompatibilities include with cyclizine, glycopyrronium, haloperidol, levomepromazine, midazolam and morphine.
- Available as: Injection 30mg/ml. Injection contains ethanol as an excipient.
- Oral 10mg tablets and injection 10mg/ml no longer available in the UK (discontinued early 2013 due to lack of demand).

Evidence: [1, 143-153]

Lactulose

Use:
- Constipation, faecal incontinence related to constipation
- Hepatic encephalopathy and coma.

Dose:
Constipation:
By mouth: initial dose twice daily then adjusted to suit patient
- **Neonate:** 2.5 ml/dose twice a day
- **Child 1 month to 1 year:** 2.5 ml/dose 1-3 times daily,
- **Child 1year to 5 years:** 5 ml/dose 1-3 times daily,
- **Child 5-10 years:** 10 ml/dose 1-3 times daily,
- **Child 10-18 years:** 15 ml/ dose 1-3 times daily.

Hepatic encephalopathy:
- **Child 12-18 years:** use 30-50 ml three times daily as initial dose. Adjust dose to produce 2-3 soft stools per day.

Notes:
- Licensed for constipation in all age groups. Not licensed for hepatic encephalopathy in children.
- Side effects may cause nausea and flatus, with colic especially at high doses. Initial flatulence usually settles after a few days.
- Precautions and contraindications; Galactosaemia, intestinal obstruction. Caution in lactose intolerance.
- Use is limited as macrogols are often better in palliative care. However the volume per dose of macrogols is 5-10 times greater than lactulose and may not be tolerated in some patients.
- Sickly taste.
- Onset of action can take 36-48 hours.
- May be taken with water and other drinks.
- Relatively ineffective in opioid induced constipation: need a stimulant.
- 15 ml/ day is 14kcal so unlikely to affect diabetics.
- Does not irritate or directly interfere with gut mucosa.
- Available as oral solution 10 g/ 15 ml. Cheaper than Movicol (macrogol).

Evidence: [1, 2, 5, 6, 43, 154-156]

Lansoprazole

Uses:
- Gastro-oesophageal reflux disease; erosive oesophagitis; prevention and treatment of NSAID gastric and oesophageal irritation; treatment of duodenal and gastric ulcer.

Dose and routes:
Oral
- **Child body weight <30kg:** 0.5-1mg/kg with maximum 15mg once daily in the morning
- **Child body weight>30kg:** 15-30mg once daily in the morning

Notes:
- Lansoprazole is not licensed in the UK for infants, children or adolescents. Lansoprazole is however licensed in the US for use from 1 year of age. Exact doses limited by available formulations.
- Lansoprazole is a gastric proton pump inhibitor. It inhibits the final stage of gastric acid formation by inhibiting the activity of H+/K+ ATPase of the parietal cells in the stomach. The inhibition is dose dependent and reversible, and the effect applies to both basal and stimulated secretion of gastric acid.
- For optimal effect, the single daily dose is best taken in the morning.
- Lansoprazole should be taken at least 30 minutes before food as intake with food slows down the absorption and decreases the bioavailability.
- The dose may be increased if symptoms do not fully resolve (consider increasing the single daily dose or BD dosing).
- Studies in infants and children indicate they appear to need a higher mg/kg dose than adults to achieve therapeutic acid suppression.
- There is some anecdotal experience that Lansoprazole FasTabs may be halved to give a 7.5mg dose
- No dose adjustment is needed in patients with renal impairment. Reduction of dose (50%) is recommended in patients with moderate to severe hepatic impairment.
- Hypomagnesaemia may develop with prolonged use
- Common adverse effects (>1 in 100 to <1 in 10): headache, dizziness; nausea; diarrhoea; stomach pain; constipation; vomiting; flatulence; dry mouth, pharyngitis; increase in liver enzyme levels; urticaria; itching, rash
- Lansoprazole may interfere with absorption of drugs where gastric pH is critical to its bioavailability (e.g. atazanavir, itraconazole); may cause increase in digoxin levels and increase in plasma concentration of drugs metabolised by CYP3A4 (e.g. theophylline and tacrolimus). Drugs which inhibit or induce CYP2C19 or CYP3A4 may affect the plasma concentration of lansoprazole.

Sucralfate and antacids may decrease the bioavailability of lansoprazole.
- Capsules: Capsules should be swallowed whole with liquid. For patients with difficulty swallowing; studies and clinical practice suggest that the capsules may be opened and the granules mixed with a small amount of water, apple/tomato juice or sprinkled onto a small amount of soft food (e.g. yoghurt, apple puree) to ease administration.
- FasTabs: Place on the tongue and gently suck. The FasTab rapidly disperses in the mouth releasing gastro-resistant microgranules which are the swallowed. FasTabs can be swallowed whole with water or mixed with a small amount of water if preferred. FasTabs contain lactose and aspartame and should be used with caution in known PKU patients.
- For administration via a NG or gastrostomy tube, lansoprazole FasTabs can be dispersed in 10ml water and administered via an 8Fr NG tube without blockage. For smaller bore tubes, dissolve the contents of a lansoprazole capsule in 8.4% sodium bicarbonate before administration. If the tube becomes blocked, use sodium bicarbonate to dissolve any enteric coated granules lodged in the tube. Lansoprazole less likely than omeprazole MUPS to cause blockage of small bore tubes.
- Available as 15mg and 30mg capsules and 15mg and 30mg FasTabs (Zoton[R])

Evidence: [1, 2, 12, 157-170]

Levomepromazine

Use
- Broad spectrum antiemetic where cause is unclear, or where probably multifactorial.
- Second line if a specific antiemetic fails.
- Severe pain unresponsive to other measures - may be of benefit in a very distressed patient.
- Sedation for terminal agitation, particularly in end of life care.

Dose and routes
Used as antiemetic
By mouth:
- **Child 2–12 years:** initial dose 50-100microgram/kg given once or twice daily. This dose may be increased as necessary and as tolerated not to exceed 1 mg/kg/dose (or maximum of 25mg/dose) given once or twice daily.,
- **Child 12-18 years:** initial dose 3mg once or twice daily. This dose may be increased as necessary and as tolerated to a maximum of 25 mg once or twice daily.

By continuous IV or SC infusion over 24 hours:
- **Child 1 month–12 years:** initial dose of 100microgram/kg/24 hours increasing as necessary to a maximum of 400microgram/kg/24 hours. Maximum 25 mg/24 hours,
- **Child 12–18 years:** initial dose of 5 mg/24 hours increasing as necessary to a maximum of 25 mg/24 hours.

Used for sedation
By SC infusion over 24 hours:
- **Child 1 year–12 years:** initial dose of 350microgram/kg/24 hours (maximum initial dose 12.5 mg), increasing as necessary up to 3 mg/kg/24 hours,
- **Child 12–18 years:** initial dose of 12.5 mg/24 hours increasing as necessary up to 200 mg/24 hours.

Analgesia
- In adults stat dose 12.5 mg/dose by mouth or SC. Titrate dose according to response; usual maximum daily dose in adults is 100 mg SC or 200 mg by mouth.

Notes:
- Licensed for use in children with terminal illness for the relief of pain and accompanying anxiety and distress
- A low dose is often effective as antiemetic. Titrate up as necessary. Higher doses are very sedative and this may limit dose increases.

- If the child is not stable on high dosage for nausea and vomiting, reconsider cause and combine with other agents.
- Some experience in adults with low dose used buccally as antiemetic (e.g. 1.5mg three times daily as needed).
- Can cause hypotension particularly with higher doses. Somnolence and asthenia are frequent side effects.
- Levomepromazine and its non-hydroxylated metabolites are reported to be potent inhibitors of cytochrome P450 2D6. Co-administration of levomepromazine and drugs primarily metabolised by the cytochrome P450 2D6 enzyme system may result in increased plasma concentrations of these drugs that could increase or prolong both their therapeutic and/or adverse effects.
- Avoid, or use with caution, in patients with liver dysfunction or cardiac disease.
- Tablets may be halved or quartered to obtain smaller doses. Tablets/segments may be dispersed in water for administration via an enteral feeding tube.
- For SC infusion dilute with sodium chloride 0.9%. Water for injection may also be used. The SC dose is considered to be twice as potent as that administered orally.
- Available as: tablets (25mg) and injection (25mg/mL). A 6mg tablet is also available via specialist importation companies. An extemporaneous oral solution may be prepared.

Evidence: [1, 2, 5, 12, 171-173] CC, EA

Lidocaine (Lignocaine) patch

Use
- Localised neuropathic pain

Dose and routes
Topical:
- **Child 3 - 18 years:** apply 1 -2 plasters to affected area(s). Apply plaster once daily for 12 hours followed by 12 hour plaster free period (to help reduce risk of skin reactions)

- **Adult 18 years or above:** up to 3 plasters to affected area(s). Apply plaster once daily for 12 hours followed by 12 hour plaster free period (to help reduce the risk of skin reactions).

Notes:
- Not licensed for use in children or adolescents under 18 years.
- Cut plaster to size and shape of painful area. Do NOT use on broken or damaged skin or near the eyes.
- When lidocaine 5% medicated plaster is used according to the maximum recommended dose (3 plasters applied simultaneously for 12 hours) about $3 \pm 2\%$ of the total applied lidocaine dose is systemically available and similar for single and multiple administrations.
- Maximum recommended number of patches in adults currently is 3 per application.
- The plaster contains propylene glycol which may cause skin irritation. It also contains methyl parahydroxybenzoate and propyl parahydroxybenzoate which may cause allergic reactions (possibly delayed). Approximately 16% of patients can be expected to experience adverse reactions. These are localised reactions due to the nature of the medicinal product.
- The plaster should be used with caution in patients with severe cardiac impairment, severe renal impairment or severe hepatic impairment.
- A recent analysis by anatomic site of patch placement suggests that application to the head was tolerated less well compared with the trunk and extremities.
- Doses extrapolated from BNF 2012 March.
- Available as 700 mg/medicated plaster (5% w/v lidocaine).

Evidence: [1, 5, 174-179] CC, EA

Lomotil® (co-phenotrope)

Use:
- Diarrhoea from non-infectious cause.

Dose and routes
By mouth:
- **Child 2–4 years:** half tablet 3 times daily,
- **Child 4–9 years:** 1 tablet 3 times daily,
- **Child 9–12 years:** 1 tablet 4 times daily,
- **Child 12–16 years:** 2 tablets 3 times daily,
- **Child 16–18 years:** initially 4 tablets then 2 tablets 4 times daily.

Notes:
- Not licensed for use in children < 4 years.
- Tablets may be crushed. For administration via an enteral feeding tube, tablets may be crushed and dispersed in water immediately before use. Young children are particularly susceptible to overdosage and symptoms may be delayed and observation is needed for at least 48 hours after ingestion. Overdose can be difficult to manage with a mixed picture of opioid and atropine poisoning. Further, the presence of subclinical doses of atropine may give rise to atropine side-effects in susceptible individuals.
- Available only as tablets Co-Phenotrope (2.5 mg diphenoxylate hydrochloride and 25 microgram atropine sulphate).

Evidence: [1, 2, 180-182]

Loperamide

Use:
- Diarrhoea from non-infectious cause.
- Faecal incontinence

Dose and routes
By mouth:
- **Child 1 month–1 year:** initial dose of 100microgram/kg twice daily given 30 minutes before feeds. Increase as necessary up to a maximum of 2 mg/kg/day given in divided doses,
- **Child 1–12 years:** initial dose of 100microgram/kg (maximum single dose 2 mg) 3-4 times daily. Increase as necessary up to a maximum of 1.25 mg/kg/day given in divided doses (maximum 16 mg/day),
- **Child 12–18 years:** initial dose of 2 mg 2-4 times daily. Increase as necessary up to a maximum of 16 mg/day given in divided doses.

Notes:
- Not licensed for use in children with chronic diarrhoea.
- Capsules not licensed for use in children < 8 years.
- Syrup not licensed for use in children < 4 years.
- Common side effects: constipation, nausea, flatulence
- As an antidiarrhoeal, loperamide is about 50x more potent than codeine. It is longer acting; maximum therapeutic impact may not be seen for 16-24 hours.
- For NG or gastrostomy administration: Use the liquid preparation undiluted. Flush well after dosing. Alternatively, the tablets can be used without risk of blockage, although efficacy is unknown. Jejunal administration will not affect the therapeutic response to loperamide. However, owing to the potential osmotic effect of the liquid preparation, it may be appropriate to further dilute the dose with water immediately prior to administration.
- Available as tablets (2 mg), orodispersible tablets (2mg) and oral syrup (1 mg/5 mL).

Evidence: [1, 2, 12, 183-185]

Lorazepam

Use
- Background anxiety.
- Agitation and distress.
- Adjuvant in cerebral irritation.
- Background management of dyspnoea.
- Muscle spasm.
- Status epilepticus.

Dose and routes for all indications except status epilepticus:
By mouth:
- **Child < 2 years:** 25 microgram/kg 2–3 times daily,
- **Child 2–5 years:** 500microgram 2–3 times daily,
- **Child 6–10 years:** 750micorgram 3 times daily,
- **Child 11–14 years:** 1 mg 3 times daily,
- **Child 15–18 years:** 1–2 mg 3 times daily.

Sublingual:
- **Children of all ages**: 25 micrograms/kg as a single dose. Increase to 50 microgram/kg (maximum 1 mg/dose) if necessary.
- **Usual adult dose**: 500 microgram – 1 mg as a single dose, repeat as required.

Notes
- Not licensed for use in children for these indications other than status epilepticus.
- Tablets licensed in children over 5 years for premedication, injection not licensed in children less than 12 years except for treatment of status epilepticus
- Potency in the order of 10 times that of diazepam per mg as anxiolytic / sedative.
- Well absorbed sublingually with rapid onset of effect. There may however be variable absorption by this route with further variation possible depending on the formulation used; fast effect.

- Specific sublingual tablets are not available in the UK but generic lorazepam tablets (specifically Genus, PVL or TEVA brands) do dissolve in the mouth to be given sublingually.
- Tablets may be dispersed in water for administration via an enteral feeding tube.
- May cause drowsiness and respiratory depression if given in large doses.
- Caution in renal and hepatic failure.
- Available as tablets (1 mg, 2.5 mg) and injection (4 mg in 1mL).

Evidence: [2, 5, 12, 125, 186] CC, EA

Macrogols: Movicol®

Use
- Constipation.
- Faecal impaction.
- Suitable for opioid-induced constipation.

Dose and routes **(Movicol® paediatric for those less than 12 years of age; Movicol(R) from 12 years of age)**
By mouth for constipation or prevention of faecal impaction:
- **Child under 1 year:** ½-1 sachet daily,
- **Child 1–6 years:** 1 sachet daily (adjust dose according to response; maximum 4 sachets daily),
- **Child 6–12 years:** 2 sachets daily (adjust dose according to response; maximum 4 sachets daily),
- **Child 12–18 years:** 1–3 sachets daily of **adult** Movicol®.

By mouth for faecal impaction:
- **Child under 1 year:** ½-1 sachet daily,
- **Child 1–5 years:** 2 sachets on first day and increase by 2 sachets every 2 days (maximum 8 sachets daily). Treat until impaction resolved then switch to maintenance laxative therapy,

- **Child 5–12 years:** 4 sachets on first day and increase by 2 sachets every 2 days (maximum 12 sachets daily). Treat until impaction resolved then switch to maintenance laxative therapy,
- **Child 12–18 years:** 4 sachets daily of adult Movicol® then increase by 2 sachets daily to a maximum of 8 adult Movicol sachets daily. Total daily dose should be drunk within a 6 hour period. After disimpaction switch to maintenance laxative therapy.

Notes
- Not licensed for use in children < 5 years with faecal impaction and < 2 years with chronic constipation.
- Need to maintain hydration. Caution if fluid or electrolyte disturbance.
- Caution with high doses in those with impaired gag reflex, reflux oesophagitis or impaired consciousness
- Mix powder with water: Movicol® paediatric 60mL per sachet and adult Movicol® 125 mL per sachet.
- For administration via a feeding tube: Dissolve the powder in water as directed and flush down the feeding tube. Flush well after dosing.
- Macrogol oral powder is also available as alternative brands Laxido Orange(R), Molaxole(R) but these preparations are licensed for use only from 12 years of age

Evidence: [1, 2, 12, 155, 187, 188]

Melatonin

Use:
- Sleep disturbance due to disruption of circadian rhythm (*not* anxiolytic).

Dose and routes
By mouth:
- **Child 1 month-18 years:** initial dose 2–3 mg, increasing every 1–2 weeks dependent on effectiveness up to maximum 12 mg daily.

Notes:
- Not licensed for use in children.
- Specialist use only.
- Some prescribers use a combination of immediate release and m/r tablets to optimise sleep patterns.
- Immediate release capsules may be opened and the contents sprinkled on cold food if preferred. Sustained release capsules may also be opened but the contents should not be chewed.
- Only licensed formulation in the UK is 2 mg m/r tablets (Circadin). Various unlicensed formulations, including an immediate release preparation are available from 'specials' manufacturers or specialist importing companies.

Evidence: [1, 2, 189-206] CC

Methadone

(WARNING: requires additional training for dosing)
Use:
- Major opioid used for moderate to severe pain, particularly neuropathic pain and pain poorly responsive to other opioids.
- Not normally used as first line analgesia.

<u>Caution:</u>

Methadone should only be commenced by practitioners experienced in its use.
This is due to wide inter-interindividual variation in response, very variable conversion ratios with other opioids, complex pharmacokinetics and a long half life.
Initial close monitoring is particularly important.

Dose and routes

In opioid naïve children
By mouth, SC and IV injection:
- **Child 1-12 years:** 100-200 micrograms / kg every 4 hours for first 2-3 doses then every 6-12 hours (maximum dose 5 mg/ dose initially)
- Methadone has a long and variable half-life with potential to cause sedation, respiratory depression and even death from secondary peak phenomenon.
- Titration of methadone dosing must be done under close clinical observation of the patient particularly in the first few days. Due to large volume of distribution, higher doses are required for the first few days whilst body tissues become saturated. Once saturation is complete, a smaller dose is sufficient. Continuing on initial daily dose is likely to result in sedation within a few days, possible respiratory depression and even death.
- **Doses may need to be reduced by up to 50% 2-3 days after the initial, effective dose has been found to prevent adverse effects** due to tissue saturation and methadone accumulation. From then on increments in methadone dosing should be approximately at weekly intervals with a maximum increase of 50% (experienced practitioners may increase more frequently).
- Continued clinical reassessment is required to avoid toxicity as the time to reach steady state concentration following a change in dosing may be up to 12 days.
- For breakthrough pain, we would recommend using a short half-life opioid.

In opioid substitution/ rotation or switch
Caution:

Substitution, rotation or switch to methadone is a specialist skill and should only be undertaken in close collaboration with practitioners experienced in its use. There is a risk of unexpected death through overdose.

Equianalgesic doses:

> Dose conversion ratios from other opioids are not static but are a function of previous opioid exposure, and are highly variable.
>
> Published tables of equianalgesic doses of opioids, established in healthy non-opioid tolerant individuals, indicate that methadone is 1–2 times as potent as morphine in single dose studies, but in individuals on long-term (and high dose) morphine, methadone is closer to 10 times as potent as morphine; it can be 30 times more potent or occasionally even more. The potency ratio tends to increase as the dose of morphine increases. If considering methadone, thought should be given to the potential difficulty of subsequently switching from methadone to another opioid.
>
> Other opioids should be considered first if switching from morphine due to unacceptable effects or inadequate analgesia.
>
> Consultation with a pain clinic or specialist palliative-care service is advised
>
> Ref [4]

In adults there are several protocols for opioid rotation to methadone which are not evidence based in paediatrics.

- In one approach, previous opioid therapy is completely stopped before starting a fixed dose of methadone at variable dose intervals.

- The other approach incorporates a transition period where the dose of the former opioid is reduced and partially replaced by methadone which is then titrated upwards.

It can be difficult to convert a short-half-life opioid to a methadone equivalent dose and vice versa. Current practice is usually to admit to a specialist inpatient unit for 5-6 days of regular treatment or titrate orally at home with close supervision.

Converting oral methadone to SC/IV or CSCI/CIVI methadone
- Approximate dose ratios for switching between oral dosage and parenteral / subcutaneous form 2:1 (oral: parenteral).
- Calculate the total daily dose of oral methadone and halve it (50%). This will be the 24hour parenteral / subcutaneous methadone dose.
- Seek specialist guidance if mixing with any other drug [207].
- If CSCI methadone causes a skin reaction, consider doubling the dilution and changing the syringe every 12 hours.
- Administer IV methadone slowly over 3-5 minutes.

Notes:
- Not licensed for use in children.
- Data on methadone in paediatric patients is limited; known to have wide inter-individual pharmacokinetic variation.
- Use methadone with caution, as methadone's effect on respiration lasts longer than analgesic effects.
- Side effects include: nausea, vomiting, constipation, dry mouth, biliary spasm, respiratory depression, drowsiness, muscle rigidity, hypotension, bradycardia, tachycardia, palpitation, oedema, postural hypotension, hallucinations, vertigo, euphoria, dysphoria, dependence, confusion, urinary retention, ureteric spasm and hypothermia.
- Following concerns regarding methadone and sudden death from prolongation of QT interval or torsade de pointes (especially at high doses) it is recommended that patients have an ECG prior to initiation of treatment and regularly whilst on methadone, particularly if they have any risk factors or are having intravenous treatment of methadone.
- Opioid antagonists naloxone and naltrexone will precipitate an acute withdrawal syndrome in methadone dependent individuals. Naloxone will also antagonize the analgesic, CNS and respiratory depressant effects of methadone.

- Methadone has the potential for a number of significant drug interactions. Drugs that induce cytochrome P450 3A4 enzymes (e.g. carbamazepine, phenobarbital, phenytoin, rifampicin, some HIV drugs) will increase the rate of metabolism of methadone and potentially lead to reduced serum levels. Drugs that inhibit the system (e.g. amitriptyline, ciprofloxacin, fluconazole) may lead to increased serum levels of methadone.
- Renal impairment: if severe (i.e. GFR <10 ml/min or serum creatinine >700 micromole/l) –reduce methadone dose by 50% and titrate according to response. Significant accumulation is not likely in renal failure, as elimination is primarily via the liver.
- As methadone has a long half-life, infusion of naloxone may be required to treat opioid overdose.
- Available as: linctus (2 mg/5 mL), mixture (1 mg/mL), solution (1 mg/mL, 5 mg/ml, 10 mg/mL, and 20 mg/mL), tablets (5 mg), and injection (10 mg/mL).

Evidence: [1, 2, 4, 5, 29, 43, 207-219]

Methylnaltrexone

Use:
- Opioid-induced constipation when the response to other laxatives alone is inadequate, and other relevant factors have been / are being addressed.

Dose and routes
SC (usual route) or IV bolus:
- **Child 1month– 12 years:** 0.15mg/kg (maximum 8mg) as a single dose
- **Child >12 years: with weight 38-61kg:** 8mg as a single dose
- **Child >12 years: with weight 62-114kg:** 12mg as a single dose
- **Child >12 years:** but with body weight less than 38kg, use 0.15mg/kg

A single dose may be sufficient. However repeat doses may be given with a usual administration schedule of a single dose every other day. Doses may be given with longer intervals, as per clinical need. Patients may receive 2 consecutive doses (24 hours apart) only when there has been no response (no bowel movement) to the dose on the preceding day. (1/3 to ½ of patients given methylnaltrexone have a bowel movement within 4 hours, without loss of analgesia).

Notes:
- μ-opioid receptor antagonist that acts exclusively in the peripheral tissues including the GI tract (increasing bowel movement and gastric emptying) and does not affect the analgesic effects of opioids.
- Not licensed for use in children or adolescents less than 18 years.
- Not licensed for IV administration – usual route is SC
- Methylnaltrexone is contraindicated in cases of known or suspected bowel obstruction.
- The onset of effect may be within 15-60 minutes.
- Common side-effects include abdominal pain/colic, diarrhoea, flatulence and nausea.
- If administered by SC injection rotate the site of injection. Do not inject into areas where the skin is tender, bruised, red or hard.
- Constipation in palliative care is usually multifactorial and other laxatives are often required in addition.
- Reduce dose by 50% in severe renal impairment.
- Does not cross blood brain barrier.
- Available as single use vial 12mg/0.6ml solution for SC injection (Relistor[(R)])

Evidence: [1, 220-226]

Metoclopramide

To minimise the risk of neurological side effects associated with metoclopramide, the EMA in 2013 issued the following recommendations: **(NB use of metoclopramide in palliative care was excluded from these recommendations HOWEVER caution should be exercised nevertheless)**.

Use of metoclopramide is contraindicated in children younger than 1 year.

In children aged 1-18 years, metoclopramide should only be used as a second-line option for prevention of delayed chemotherapy-induced nausea and vomiting and for treatment of established postoperative nausea and vomiting.

Metoclopramide should only be prescribed for short term use (up to 5 days).

Use
- Antiemetic if vomiting caused by gastric compression or hepatic disease.
- Prokinetic for slow transit time (not in complete obstruction or with anticholinergics).
- Hiccups.

Dose and routes
By mouth, IM injection, or IV injection (over at least 3 minutes):
- **Neonate:** 100 microgram/kg every 6–8 hours (by mouth or IV only),
- **Child 1 month–1 year and body weight up to 10 kg:** 100 microgram/kg (maximum. 1 mg/dose) twice daily,
- **Child 1–18 years: 100-150microgram/kg** repeated up to 3 times daily. The maximum dose in 24 hours is 500microgram/kg (maximum 10mg/dose).

If preferred the appropriate total daily dose may be administered as a continuous SC or IV infusion over 24 hours.

Notes:
- Not licensed for use in infants less than 1 year of age.
- Not licensed for continuous IV or SC infusion.
- Metoclopramide can induce acute dystonic reactions such as facial and skeletal muscle spasms and oculogyric crises; children (especially girls, young women, and those under 10 kg) are particularly susceptible. With metoclopramide, dystonic effects usually occur shortly after starting treatment and subside within 24 hours of stopping it.
- Intravenous doses should be administered as a slow bolus over at least 3 minutes to reduce the risk of adverse effects.
- Oral liquid formulations should be given via a graduated oral syringe to ensure dose accuracy in children. The oral liquid may be administered via an enteral feeding tube.
- Can be irritant on SC administration; dilute well in 0.9% NaCl
- Available as: tablets (10 mg), oral solution (5 mg/5 mL) and injection (5 mg/mL).

Evidence: [1-3, 12, 43, 45, 47, 50, 85, 87, 227-231]

Metronidazole topically

Use:
- Odour associated with fungating wounds or lesions.

Dose and routes
By topical application:
- Apply to clean wound 1–2 times daily and cover with non-adherent dressing.
- Cavities: smear gel on paraffin gauze and pack loosely.

Notes:
- Anabact® not licensed for use in children < 12 years.
- Metrogel® not licensed for use with children.
- Available as: gel (Anabact® 0.75%, Metrogel® 0.75%)

Evidence: [1, 2, 232]

Miconazole oral gel

Use:
- Oral and intestinal fungal infection.

Dose and routes
By mouth:
Prevention and treatment of oral candidiasis
- **Neonate:** 1 mL 2-4 times a day smeared around inside of mouth after feeds,
- **Child 1 month–2 years:** 1.25 mL 4 times daily smeared around inside of mouth after food,
- **Child 2–18 years:** 2.5 mL 4 times daily after meals; retain near lesions before swallowing (orthodontic appliances should be removed at night and brushed with gel).

Prevention and treatment of intestinal candidiasis
- Child 4 months – 18 years: 5 mg/kg 4 times daily; max. 250 mg (10 mL) 4 times daily.

Notes:
- Use after food and retain near lesions before swallowing.
- Treatment should be continued for 7 days after lesions have healed.
- Not licensed for use in children under 4 months or during the first 5-6 months of life of an infant born preterm.
- Infants and babies: The gel should not be applied to the back of the throat due to possible choking. The gel should not be swallowed immediately, but kept in the mouth as long as possible.
- Available as: oral gel (24 mg/mL in 15 g and 80 g tube).
- A buccal tablet of miconazole is now available. Indicated for the treatment of oropharyngeal candidiasis in immunocompromised adults, Loramyc[R] 50mg muco-adhesive buccal tablets should be applied to the upper gum just above the incisor tooth once daily for 7-14 days. Currently no experience in children but may be an option for adolescents.

Evidence: [2, 233]

Midazolam

Use:
- Status epilepticus and terminal seizure control.
- Breakthrough anxiety, e.g. panic attacks.
- Adjuvant for pain of cerebral irritation.
- Anxiety induced dyspnoea.
- Agitation at end of life.

Dose and routes

By oral or gastrostomy administration for *anxiety or sedation*:
- **1month – 18 years:** 500microgram/kg (maximum 20mg) as a single dose.

Buccal doses for **acute anxiety:**
- **Any age:** 100 microgram/kg as a single dose (maximum initial dose 5 mg).

By SC or IV infusion over 24 hours *for anxiety*:
- Dosages of 30-50% of terminal seizure control dose can be used to control anxiety, terminal agitation and terminal breathlessness.

Buccal doses for *status epilepticus:*
- **Neonate:** 300 microgram/kg as a single dose, repeated once if necessary,
- **Child 1–3 months:** 300 microgram/kg (maximum initial dose 2.5 mg), repeated once if necessary,
- **Child 3 months–1 year:** 2.5 mg, repeated once if necessary,
- **Child 1–5 years:** 5 mg, repeated once if necessary,
- **Child 5–10 years:** 7.5 mg, repeated once if necessary,
- **Child 10–18 years:** 10 mg, repeated once if necessary.

By buccal or intranasal administration for *status epilepticus*, should wait 10 minutes before repeating dose.

NB / In single dose for seizures, midazolam is twice as potent as rectal diazepam. For patients who usually receive rectal diazepam for management of status, consider an initial dose of buccal midazolam that is 50% of their usual rectal diazepam dose to minimise the risk of respiratory depression

By SC or IV infusion over 24 hours for ***terminal seizure control***:
- **Neonate** (*seizure control*): 150 microgram/kg IV loading dose followed by a continuous IV infusion of 60 microgram/kg/hour. Dose can be increased by 60 microgram/kg/hour every 15 minutes until seizure controlled (maximum dose 300 microgram/kg/hour),
- **Child 1 month – 18 years:** Initial dose 50 microgram/kg/hour increasing up to 300 microgram/kg/hour (maximum 100 mg/24 hours or 150 mg/24 hours in specialist units).

Notes
- Buccal (Buccolam oromucosal solution) midazolam is not licensed for use in infants less than 3 months of age. Midazolam injection is not licensed for use in seizure control or anxiety.
- In single dose for sedation midazolam is 3 times as potent as diazepam, and in epilepsy it is twice as potent as diazepam. (Diazepam gains in potency with repeated dosing because of prolonged half life).
- Recommended SC/IV doses vary enormously in the literature. If in doubt, start at the lowest recommended dose and titrate rapidly.
- Onset of action by buccal and intranasal route 5-15 minutes. For buccal administration, if possible, divide the dose so half is given into one cheek and the remaining half into the other cheek.
- Onset of action by oral or gastrostomy route 10-30 minutes.
- Onset of action by IV route 2-3 minutes; SC route 5-10 minutes.
- Midazolam has a short half life.
- High doses can lead to paradoxical agitation.
- Available as: oral solution (2.5 mg/mL unlicensed), buccal liquid (5 mg/mL Buccolam[R]), and injection 1mg/mL, 2mg/mL, 5mg/mL). Other oral and buccal liquids (e.g. Epistatus[R] 10mg/ml) are also available from 'specials' manufacturers or specialist importing companies (unlicensed). NOTE The buccal and oral formulations available may differ in strength – take care with prescribing.

Evidence: [2, 6, 78, 79, 81, 234-239]

Morphine

Use:
- Major opioid (WHO step 2).
- First line oral opioid for pain.
- Dyspnoea.
- Cough suppressant

Dose and routes:
Opioid naïve patient: Use the following start doses. (The maximum dose stated applies to **starting** dose only).
Opioid conversion: Convert using OME (Oral Morphine Equivalent) from previous opioid.

By mouth or by rectum
- **Child 1–3 months:** initially 50 micrograms/kg every 4 hours, adjusted according to response
- **Child 3–6 months:** 100 micrograms/kg every 4 hours, adjusted according to response
- **Child 6–12 months:** 200 micrograms/kg every 4 hours, adjusted according to response
- **Child 1–12 years:** initially 200–300 micrograms/kg (initial maximum 5-10 mg) every 4 hours, adjusted according to response
- **Child 12–18 years:** initially 5–10 mg every 4 hours, adjusted according to response

By single SC injection or IV injection (over at least 5 minutes):
- **Neonate:** initially 25 micrograms/kg every 6 hours adjusted to response,
- **Child 1-6 months:** initially 50-100 micrograms/kg every 6 hours adjusted to response,
- **Child 6 months-2 years:** initially 100 micrograms/kg every 4 hours adjusted to response,

- **Child 2-12 years:** initially 100 micrograms/kg every 4 hours adjusted to response, maximum initial dose of 2.5mg,
- **Child 12-18 years:** initially 2.5-5 mg every 4 hours adjusted to response (maximum initial dose of 20 mg/24 hours).

By continuous SC or IV infusion:
- **Neonate:** 5 micrograms/kg/hour adjusted according to response,
- **Child 1-6 months:** 10 micrograms/kg/hour adjusted according to response,
- **Child 6 months - 18 years:** 20 micrograms/kg/hour (maximum initial dose of 20 mg/24 hours) adjusted according to response.

Parenteral dose: 30-50% of oral dose if converting from oral dose of morphine

Dyspnoea
30-50% of the dose used for pain.

Notes:
- *Oramorph®* solution not licensed for use in children under 1 year; *Oramorph®* unit dose vials not licensed for use in children under 6 years; *Sevredol®* tablets not licensed for use in children under 3 years; *Filnarine®* SR tablets not licensed for use in children under 6 years; *MST Continus®* preparations licensed to treat children with cancer pain (age-range not specified by manufacturer); *MXL®* capsules not licensed for use in children under 1 year; suppositories not licensed for use in children Caution in renal or hepatic impairment.
- Where opioid substitution or rotation is to morphine: use oral morphine equivalency (OME).
- Particular side effects include urinary retention and pruritus in paediatric setting, in addition to the well recognised constipation, nausea and vomiting.
- Morphine toxicity often presents as myoclonic twitching.
- Rectal route should be avoided if possible, and usually contraindicated in children with low platelets and/or neutropenia.

- In an emergency, when oral intake not appropriate, MST tablets can be administered rectally.
- Administration via enteral feeding tubes: For immediate pain relief use oral solution; no further dilution is necessary. The tube must be flushed well following dosing to ensure that the total dose is delivered. For sustained pain relief, use MST Continus sachets, dispersed in at least 10 mL of water. Flush the tube well following dosing to ensure that the total dose is delivered. Note that any granules left in the tube will break down over a period of time and a bolus of morphine will be delivered when the tube is next flushed; this has resulted in a reported fatality. Ensure that dose prescribed can be administered using whole sachets when possible. Use of Zomorph capsules opened to release the granules should be done with caution in children due to issues with dose accuracy and the granules should only be administered via an adult size gastrostomy.

Available as: (all Schedule 2 CD except oral solution of strength 10mg in 5ml)
- Tablets (10 mg, 20 mg, 50 mg)
- Oral solution (10 mg/5 mL (POM), 100 mg/5 mL)
- Modified release tablets and capsules 12 hourly (5 mg, 10 mg, 15 mg, 30 mg, 60 mg, 100 mg, 200 mg).
- Modified release suspension 12 hourly (20 mg, 30 mg, 60 mg, 100 mg, 200 mg).
- Modified release capsules 24 hourly (30 mg, 60 mg, 120 mg, 150mg, 200 mg).
- Suppositories (10 mg, 15 mg, 30 mg).
- Injection (10 mg/mL, 15 mg/mL, 20 mg/mL and 30 mg/mL)

Evidence: [1-3, 6, 12, 27, 29, 77, 100, 132, 240-259]

Nabilone

Use:
- Nausea and vomiting caused by cytotoxic chemotherapy (not first or second line therapy).
- For unresponsive nausea and vomiting to conventional antiemetics.

Dose and routes
By mouth:
- **Adult dose:** 1–2 mg twice a day (maximum dose 6 mg/day in 2-3 divided doses)

Notes:
- Not licensed for use in children.
- Nabilone is a synthetic cannabinoid.
- Individual variation requiring close medical supervision on commencement and dose adjustments.
- The effects of Nabilone may persist for a variable and unpredictable period of time following its oral administration. Adverse psychiatric reactions can persist for 48 to 72 hours following cessation of treatment.
- For specialist use only.
- Available as: capsules (1 mg). Schedule 2 controlled drug.

Evidence: EA [1, 2, 5]

Naloxone

Use:
- Emergency use for reversal of opioid-induced respiratory depression or acute opioid overdose.
- Constipation when caused by opioids if methylnaltrexone not available and laxatives have been ineffective.

Dose and routes

Reversal of respiratory depression due to opioid overdose
By intravenous injection: (review diagnosis; further doses may be required if respiratory depression deteriorates)
- **Neonate:** 10 micrograms/kg; if no response give a subsequent dose of 100 microgram/kg (then review diagnosis),
- **Child 1 month-12 years:** 10 micrograms/kg; if no response give a subsequent dose of 100 microgram/kg (then review diagnosis),
- **Child 12-18 years:** 400microgram-2 mg; if no response repeat at intervals of 2-3 minutes to maximum of 10 mg total dose (then review diagnosis).

By subcutaneous or intramuscular injection only if intravenous route not feasible
- As per intravenous injection but onset slower and potentially erratic.

By continuous intravenous infusion, adjusted according to response
- **Neonate:** Rate adjusted to response (initially, rate may be set at 60% of the initial resuscitative intravenous injection dose per hour).
- **Child 1 month-18 years:** Rate adjusted to response (initially, rate may be set at 60% of the initial resuscitative intravenous injection dose per hour).
- *The initial resuscitative intravenous injection dose is that which maintained satisfactory self ventilation for at least 15 minutes.*

Opioid-induced constipation

By mouth:
- In adults the following doses have been used: total daily dose oral naloxone = 20% of morphine dose; titrate according to need; maximum single dose 5 mg.

Notes
- Potent opioid antagonist.

- Not licensed for use in children with constipation.
- Also see methylnaltrexone
- Naloxone acts within 2 minutes of IV injection and within 3-5 minutes of SC or IM injection.
- Although oral availability of naloxone is relatively low, be alert for opioid withdrawal symptoms, including recurrence of pain, at higher doses.
- Available as: injection (20 microgram/mL, 400microgram/ml, 1mg/ml).

Evidence: [2, 260, 261] EA

Naproxen

Uses:
- Non-steroidal anti-inflammatory agent analgesic; relief of symptoms in inflammatory arthritis and treatment of acute musculoskeletal syndromes.

Dose and route:
- **Child 1 month -18 years:** 5mg/kg/dose BD (maximum 1g/ day)

Doses up to 10mg/kg BD (not exceeding 1g daily) have been used in severe conditions. High doses should ideally be used only for a short period. In general, use the lowest effective dose for the shortest treatment duration possible.

Notes:
- Naproxen is licensed for use from 5 years of age for juvenile idiopathic arthritis; not licensed for use in children less than 16 years for other conditions.
- Naproxen is contraindicated in patients with a history of hypersensitivity to any NSAID or in those with a coagulation disorder.
- Use with caution in renal, cardiac or hepatic failure as this may cause a deterioration in renal function; the dose should be kept as

- low as possible and renal function monitored. Avoid use if GFR <20ml/min/1.73m2 and in those with severe hepatic or heart failure.
- Generally naproxen is regarded as combining good efficacy with a low incidence of side-effects.
- The risk of cardiovascular events secondary to NSAID use is undetermined in children. In adults COX-2 selective inhibitors, diclofenac (150mg daily) and ibuprofen (2.4g daily) are associated with an increased risk of thrombotic effects (e.g. myocardial infarction and stroke). Naproxen (in adults 1g daily) is associated with a lower thrombotic risk. The greatest risk may increase with dose and duration of exposure so the lowest effective dose should be used for the shortest possible duration of time.
- All NSAIDs are associated with GI toxicity. In adults, evidence on the relative safety of NSAIDs indicates differences in the risks of serious upper GI side-effects – piroxicam and ketorolac are associated with the highest risk; indometacin, diclofenac and naproxen are associated with intermediate risk and ibuprofen with the lowest risk. Children appear to tolerate NSAIDs better than adults and GI side-effects are less common although they do still occur and can be significant.
- Other potential side-effects include headache, dizziness, vertigo, fluid retention and hypersensitivity reactions.
- The antipyretic and anti-inflammatory actions of naproxen may reduce fever and inflammation therefore reducing their utility as diagnostic signs.
- Potential drug interactions include warfarin (increase in INR); diuretics, ACE inhibitors and angiotensin II antagonists (increased risk of compromised renal function). Naproxen is a substrate of CYP1A2 and CYP2C8/9 and can increase the plasma concentrations of methotrexate and lithium.
- Naproxen tablets may be crushed before administration and can be mixed with water for administration via a feeding tube. However, naproxen is poorly soluble in water and the tablet must be crushed to a fine powder before mixing with water to avoid tube blockage. There may be better choices of NSAID if administration via a

feeding tube is necessary. Enteric coated naproxen tablets should be swallowed whole and NOT be crushed or chewed. Naproxen should be taken with or after food.
- Available as: tablets 250mg and 500mg; enteric coated tablets 250mg and 500mg; oral suspension 25mg/ml (available only as a 'special' from specials manufacturers).

Evidence: [1, 2, 5, 12]

Nystatin

Use:
- Oral and perioral fungal infection.

Dose and routes
By mouth:
- **Neonate:** 100 000 units 4 times a day,
- **Child 1 month–18 years:** 100 000 units 4 times a day.

Notes:
- Licensed from 1 month of age for treatment. Neonates – nystatin is licensed for prophylaxis against oral candidosis at a dose of 1ml daily.
- Retain near lesions before swallowing.
- Administer after food or feeds.
- Treatment for 7 days and should be continued for 48 hours after lesions have healed.
- Available as: oral suspension 100 000 units/mL, 30 mL with pipette.

Evidence: [2, 108, 262]

Octreotide

Use:
- Bleeding from oesophageal or gastric varices.
- Nausea and vomiting.
- Intestinal obstruction.
- Intractable diarrhoea.
- Also used for hormone secreting tumours, ascites, bronchorrhoea.

Dose and routes
Bleeding from oesophageal varices
By continuous intravenous or subcutaneous infusion
- **Child 1month-18 years:** 1 microgram/kg/hour. Higher doses may be required initially. When there is no active bleeding reduce dose over 24 hours. Usual maximum dose is 50 micrograms/hour.

Nausea and vomiting, intestinal obstruction and intractable diarrhoea
By continuous intravenous or subcutaneous infusion: doses up to 1micorgram/kg/hour have been used but experience is limited. Do not stop abruptly -discontinue at a reducing rate.

Notes:
- Not licensed for use in children.
- Administration: for IV injection or infusion, dilute with sodium chloride 0.9% to a concentration of 10-50% (i.e. not less than 1:1 and not more than 1:9 by volume). For SC bolus injections, may be administered neat but this can be painful (this can be reduced if the ampoule is warmed in the hand to body temperature before injection). For SC infusion dilute with 0.9% NaCl.
- Avoid abrupt withdrawal.
- Available as: injection for SC or IV administration (50 micrograms/mL, 100 micrograms/ml, 200 micrograms/ml, 500 micrograms/mL). Also available as depot injection for IM administration every 28 days (10 mg, 20 mg and 30 mg Sandostatin Lar[R]). Recommend specialist palliative care advice.

Evidence: [2, 5, 43]

Olanzapine

Uses:
- Psychoses; delirium; agitation; nausea and vomiting; anorexia when all other treatments have failed.

Dose and route:
Oral:

Psychoses / mania
- **Child <12 years and <25kg:** initial dose 2.5mg at night,
- **Child <12 years and >25kg:** initial dose 2.5-5mg at night,
- **Child 12-18 years:** initial dose 5mg at bedtime.

Increase gradually as necessary and as tolerated to a maximum of 20mg/day given usually as a single dose at night.

Agitation/delirium
- **Child <12 years:** initial dose 1.25mg at night and PRN,
- **Child 12-18 years:** initial dose 2.5mg at night and PRN.

Increase gradually as necessary and as tolerated to maximum 10mg/day.

Nausea and vomiting; anorexia
- **Child <12 years:** initial dose 1.25mg (or 0.625mg if 2.5mg tablets can be cut into quarters) at night and PRN,
- **Child 12-18 years:** initial dose 1.25-2.5mg at night and PRN.

Dose may be increased as necessary and as tolerated to a suggested maximum of 7.5mg/day.

Notes:
- Olanzapine is not licensed for use in children and adolescents less than 18 years of age although there is general acknowledgement of 'off-label' use in adolescents for the treatment of psychosis and schizophrenia and mania associated with bipolar disorder.

- Use in the treatment of agitation/delirium, nausea and vomiting and anorexia in palliative care are all 'off-label' indications.
- Olanzapine is an atypical (second generation) antipsychotic agent and antagonist of D_1, D_2, D_3, D_4, $5\text{-}HT_{2A}$, $5HT_3$, $5HT_6$, histamine-1 and muscarinic receptors.
- Olanzapine has 5x the affinity for $5HT_2$ receptors than for D_2 receptors resulting in fewer extrapyramidal side-effects.
- Activity of olanzapine at multiple receptors is similar to methotrimeprazine and therefore it has a potential role in the treatment of nausea and vomiting refractory to standard medication.
- Use with caution in those with cardiovascular disease or epilepsy (and conditions predisposing to seizures).
- Very common (> 10% patients) adverse effects: weight gain; elevated triglyceride levels; increased appetite; sedation; increased ALT and AST levels; decreased bilirubin; increased GGT and plasma prolactin levels. Common (1-10% patients) adverse effects: elevated cholesterol levels; dry mouth.
- Rare but potentially serious adverse effects include neuroleptic malignant syndrome and neutropenia. Hyperglycaemia and sometime diabetes can occur.
- Dose titration should be slow to minimise sedation.
- A greater magnitude of weight gain and lipid and prolactin alterations have been reported in adolescents compared to adults. If prolonged use is likely, consider the monitoring of blood lipids, weight, fasting blood glucose and prolactin. Consider an ECG and BP measurement before initiation.
- Consider lower starting dose (maximum 5mg in adults) in patients with renal and/or hepatic impairment.
- Olanzapine has good oral bioavailability with peak plasma concentrations occurring within 5-8 hours. Absorption is not affected by food. Long elimination half-life of ~33 hours. Onset of actions is hours-days in delirium; days-weeks in psychoses.
- Olanzapine does not inhibit or induce the main CYP450 isoenzymes. Olanzapine is metabolised by CYP1A2 therefore drugs/substances that specifically induce or inhibit this isoenzyme

may affect the pharmacokinetics of olanzapine e.g. carbamazepine, fluvoxamine, nicotine.
- Orodispersible tablets: place in mouth where the tablet will rapidly disperse in saliva **or** disperse in a full glass of water (or other drink) immediately before administration. May be dispersed in water for administration via a NG or gastrostomy feeding tube. Some anecdotal experience that 5mg orodispersible tablets may be halved to give a 2.5mg dose. Halve immediately before administration and do not save the remaining half for a future dose
- Coated tablets: swallow whole with liquid or crushed and mixed with soft food.
- Orodispersible tablets contain aspartame and may be harmful for people with PKU.
- Coated tablets contain lactose.
- Available as: tablets 2.5mg, 5mg, 7.5mg, 10mg, 15mg, 20mg; orodispersible tablets 5mg, 10mg, 15mg, 20mg.

Evidence: [1, 2, 263-276]

Omeprazole

Use:
- Gastro-oesophageal reflux.
- Acid related dyspepsia.
- Treatment of duodenal and gastric ulcers.
- Gastrointestinal prophylaxis (e.g. with combination NSAID/steroids).

Dose and routes
By mouth:
- **Neonate:** 700 microgram/kg once daily; increase if necessary to a maximum of 1.4 mg/kg once daily (max dose: 2.8mg/kg once daily),

- **Child 1 month–2 years:** 700 microgram/kg once daily; increase if necessary to a maximum of 3 mg/kg once daily (max: 20mg once daily)
- **Child body weight 10–20 kg:** 10 mg once daily; increase if necessary to a maximum of 20 mg once daily.
- **Child body weight > 20 kg:** 20 mg once daily; increase if necessary to a maximum of 40 mg once daily.

Intravenous (by injection over 5minutes or by infusion over 20-30 minutes)
- **Child 1 month -12 years:** initially 500 micrograms/kg (max: 20 mg) once daily, increased, if necessary to 2 mg/kg (max: 40 mg) once daily,
- **Child 12-18 years:** 40 mg once daily.

Notes:
- Oral formulations are not licensed for use in children except for severe ulcerating reflux oesophagitis in children > 1 year.
- Injection not licensed for use in children under 12 years.
- Many children with life limiting conditions have gastro-oesophageal reflux disease and may need to continue with treatment long term.
- Can cause agitation.
- Occasionally associated with electrolyte disturbance.
- For oral administration tablets can be dispersed in water or with fruit juice or yoghurt. Capsules can be opened and mixed with fruit juice or yoghurt.
- Administer with care via enteral feeding tubes to minimise risk of blockage. Capsules may be opened and contents dispersed in 8.4% sodium bicarbonate for administration. Dispersible tablets disintegrate to give a dispersion of small granules. The granules settle quickly and may block fine-bore feeding tubes (less than 8Fr). .
- Available as: MUPS tablets (10 mg, 20 mg, 40 mg), capsules (10 mg, 20 mg, 40 mg), intravenous injection (40 mg) and intravenous infusion (40 mg), oral suspension available as special order (10 mg/5 mL).

Evidence: [1-3, 12, 170, 277-279]

Ondansetron

Use:
- Antiemetic, if vomiting caused by chemotherapy or radiotherapy.
- May have a use in managing opioid induced pruritus.

Dose and routes

Prevention and treatment of chemotherapy- and radiotherapy-induced nausea and vomiting
- By intravenous infusion over at least 15 minutes
Child 6 months–18 years: e*ither* 5 mg/m^2 immediately before chemotherapy (max. single dose 8 mg), then give by mouth, *or* 150 micrograms/kg immediately before chemotherapy (max. single dose 8 mg) repeated every 4 hours for 2 further doses, then give by mouth; max. total daily dose 32 mg

- By mouth following intravenous administration
Note:

Oral dosing can start 12 hours after intravenous administration

Child 6 months–18 years:

Body surface area less than 0.6 m^2 *or* body-weight 10 kg or less: 2 mg every 12 hours for up to 5 days (max. total daily dose 32 mg)

Body surface area 0.6 m^2 – 1.2m^2 or greater *or* body-weight over 10 kg: 4 mg every 12 hours for up to 5 days (max. total daily dose 32 mg)

Body surface area greater than 1.2 m^2 *or* body-weight over 40 kg: 8mg every 12 hours for up to 5 days (max. total daily dose 32 mg)

Nausea and vomiting

By mouth or slow intravenous injection over 2-5 minutes or by intravenous infusion over 15 minutes
- **Child 1-18 years:** 100-150microgram/kg/dose every 8-12 hours. Maximum single dose 4 mg.

Notes:
- Ondansetron injection is licensed for the management of chemotherapy-induced nausea and vomiting (CINV) in children aged ≥6 months, and for the prevention and treatment of post operative nausea and vomiting (PONV) in children (as a single dose) aged ≥1 month. Oral ondansetron is licensed from 6 months of age for the management of CINV but the oral formulation is not recommended for PONV in children due to a lack of data.
- Ondansetron prolongs the QT interval in a dose-dependent manner. In addition, post-marketing cases of Torsade de Pointes have been reported in patients using ondansetron. Avoid ondansetron in patients with congenital long QT syndrome. Ondansetron should be administered with caution to patients who have or may develop prolongation of QTc, including patients with electrolyte abnormalities, congestive heart failure, bradyarrhythmias or patients taking other medicinal products that lead to QT prolongation or electrolyte abnormalities.
- Hypokalaemia and hypomagnesaemia should be corrected prior to ondansetron administration.
- Repeat IV doses of ondansetron should be given no less than 4 hours apart.
- Can cause constipation and headache.
- For intravenous infusion, dilute to a concentration of 320–640 micrograms/mL with Glucose 5% or Sodium Chloride 0.9% or Ringer's Solution; give over at least 15 minutes.
- Available as: tablets (4 mg, 8 mg), oral lyophilisate (4 mg, 8 mg), oral syrup (4 mg/5 mL), injection (2 mg/mL, 2 mL and 4 mL amps).

Source: [2, 6, 44, 75, 228, 280-282]

Oxycodone

Use:
- Alternative opioid for severe pain
- Pain of all types unless opioid insensitive

Dose and routes

Opioid switch: Convert using OME (Oral Morphine Equivalent) from previous opioid.

Use the following **starting** doses in the opioid naive patient. The maximum dose stated applies to the **starting** dose only.

By mouth:
Conversion
- Oral Morphine 1.5: Oral Oxycodone 1,
- i.e. 15mg Morphine: 10mg Oxycodone
- **Child 1 - 12 months:** initial dose 50-125 micrograms/kg every 4-6 hours,
- **Child 1 - 12 years:** initial dose 125-200 micrograms/kg (maximum single dose 5 mg) every 4-6 hours,
- **Child 12-18 years:** starting dose 5 mg every 4-6 hours.
- Titrate as for morphine: Increase dose if necessary according to severity of pain.
- **m/r tablets Child 8-12 years:** initial dose 5 mg every 12 hours, increased if necessary,
- **m/r tablets Child 12-18 years:** initial dose 10 mg every 12 hours, increased if necessary.

By intravenous injection, subcutaneous injection or continuous subcutaneous infusion:
Conversion:
- Oral to IV or SC Oxycodone single bolus dose injection: Divide the oral Oxycodone dose by 1.5.
- Oral to a continuous subcutaneous infusion of Oxycodone over 24 hours: Divide the total daily dose of oral Oxycodone by 1.5.

- SC/IV Morphine to SC/IV Oxycodone ratio is approximately 1:1. i.e. use same dose.
- Reason behind odd conversion ratio is bio-availability and rounding factors for safety.

Notes:
- Opioid analgesic.
- Not licensed for use in children.
- It is important to prescribe breakthrough analgesia which is 5-10% of the total 24 hour dose given every 1 to 4 hours.
- It is moderately different from morphine in its structure, making it a candidate for opioid substitution.
- Caution in hepatic or renal impairment.
- Oxycodone injection may be given IV or SC as a bolus or by infusion. For CSCI, dilute with WFI, 0.9% sodium chloride or 5% glucose.
- Oxycodone liquid may be administered via an enteral feeding tube.
- Controlled drug schedule 2.
- Available as: tablets and capsules(5 mg, 10 mg, 20 mg), liquid (5 mg/5 ml, 10 mg/ml) and m/r tablets (5 mg, 10 mg, 15mg, 20 mg, 30mg, 40 mg, 80 mg, 120mg), injection (10 mg/ml and 50 mg/ml).

Evidence: [1, 2, 5, 12, 97, 283-287]

Oxygen

Use
- Breathlessness caused by hypoxaemia.
- Palliative care for symptom relief including recognition of potential placebo effect.

Dose and routes:
By inhalation through nasal cannula
- Flow rates of 1 – 2.5L/min adjusted according to response. This will deliver between 24 – 35% oxygen depending on the patient's

breathing pattern and other factors. Lower flow rates may be appropriate particularly for preterm neonates.

By inhalation through facemask
- Percentage inhaled oxygen is determined by the oxygen flow rate and/or type of mask. 28% oxygen is usually recommended for continuous oxygen delivery.

Notes:
- Oxygen saturations do not necessarily correlate with the severity of breathlessness. Where self-report is not possible observation of the work of breathing is a more reliable indicator of breathlessness.
- Frequent or continuous measurement of oxygen saturations may lead to an over-reliance on technical data and distract from evaluation of the child's overall comfort, symptom relief and wellbeing.
- Target oxygen saturations 92 – 96% may be appropriate in acute illness but are not necessarily appropriate for palliative care. More usual target oxygen saturations are above 92% in long-term oxygen therapy and 88-92% in children at risk of hypercapnic respiratory failure. Lower saturation levels may be tolerated in children with cyanotic congenital heart disease.
- It is important to be clear about the overall aims of oxygen treatment and realistic saturation levels for an individual child, as this will affect decisions about target oxygenation.
- In cyanotic congenital heart disease, oxygen has little effect in raising SaO2 and is not generally indicated, although the degree of polycythaemia may be reduced. Pulmonary hypertension, in the early stages, may respond to oxygen, so it may be appropriate in the palliative care setting.
- Moving air e.g. from a fan maybe equally effective in reducing the sensation of breathlessness when the child is not hypoxaemic.
- Nasal cannulae are generally preferable as they allow the child to talk and eat with minimum restrictions. However continuous nasal oxygen can cause drying of the nasal mucosa and dermatitis.
- Oxygen administration via a mask can be claustrophobic.

- The duration of supply from an oxygen cylinder will depend on the size of the cylinder and the flow rate.
- An oxygen concentrator is recommended for patients requiring more than 8 hours oxygen therapy per day.
- Liquid oxygen is more expensive but provides a longer duration of portable oxygen supply. Portable oxygen concentrators are now also available.
- If necessary, two concentrators can be Y-connected to supply very high oxygen concentrations.
- Higher concentrations of oxygen are required during air travel.
- Home oxygen order forms (HOOF) and further information available from www.bprs.co.uk/oxygen.html
- A secondary supply of oxygen for children spending a prolonged time away from home requires a second HOOF available from the above website e.g. short breaks, holiday or extended periods with other relatives

Evidence: [1, 2, 288-292]

Pamidronate (Disodium)

Use:
- Adjuvant analgesic for bone pain caused by metastatic disease.
- Adjuvant analgesic for bone pain in neurological and neuromuscular disorders, particularly due to osteopaenia or osteoporosis.
- Tumour-induced hypercalcaemia.
- Treatment of secondary osteoporosis to reduce fracture risk.

NB Evidence base is poor but growing for these uses in children. Seek specialist advice before use.

Dose and routes
For bone pain (metastatic bone disease or osteopenia); secondary osteoporosis:
An effect on pain should be seen within 2 weeks, but may need a year before definitive assessment. Continue dosing for as long as effective and tolerated or until substantial decline in performance status.

By IV infusion
- 1mg/kg as a single dose infused over 4-6 hours repeated monthly as required; concentration not exceeding 60mg in 250ml.
OR
- 1mg/kg infused over 4-6 hours on 3 consecutive days and repeated every 3 months as required; concentration not exceeding 60mg in 250ml.

For malignant hypercalcaemia: **(Seek specialist advice)**
By IV infusion
- 1 mg/kg infused over 6 hours; concentration not exceeding 60mg in 250ml. Then repeated as indicated by corrected serum calcium.

Notes:
- Not licensed for use in children. Well tolerated by children, but long term effects unknown.
- Bisphosphonates have been used for some years in adults with bone metastases. It is becoming clear that they have a role in the wider causes of bone pain seen in children, particularly with neurological conditions.
- Current guidelines suggest initial dose be given as an inpatient. Subsequent doses could be given at home, if the necessary medical and nursing support is available. May have worsening of pain at first.
- IV zoledronic acid can also be used.
- Oral risedronate and oral alendronate limited use for these indications due to poor and variable bio-availability.
- If the IV route is unavailable, bisphosphonates can be administered by CSCI over 12-24hours, together with SC hydration.
- Many bisphosphonates are available in different formulations, including oral, although absorption tends to be poor by the oral route and further reduced by food or fluids other than plain water.
- Caution: monitor renal function and electrolytes; ensure adequate hydration.
- Prolonged hypocalcaemia and hypomagnesaemia may occur with concurrent use of aminoglycoside and a bisphosphonate.

- Consider calcium and vitamin D oral supplements to minimise potential risk of hypocalcaemia for those with mainly lytic bone metastases and at risk of calcium or vitamin D deficiency (e.g. through malabsorption or lack of exposure to sunlight).
- Risk of renal impairment is increased by concurrent use with other nephrotoxic drugs.
- Risk of atypical femoral fractures, and of osteonecrosis especially of jaw if pre-existing pathology. Recommend dental check pre administration.
- Anecdotal risk of iatrogenic osteopetrosis with prolonged use (if prolonged use is likely, precede with DEXA scan and investigation of calcium metabolism).
- Available as: injection vials for infusion of various volumes, at 3mg/ml, 6mg/ml, 9mg/ml, 15mg/ml.

Evidence: CC, EA [1, 5, 293-300]

Paracetamol

Use:
- Mild to moderate pain,
- Pyrexia.

Dose:
The recommended indications and doses of paracetamol have been revised to take account of MHRA and Toxbase advice that paracetamol toxicity may occur with doses between 75-150mg/kg/day (ingestion of over 150mg/kg/day is regarded as a definite risk of toxicity).

Oral
- **Neonate 28 – 32 weeks corrected gestational age:** 20 mg/kg as a single dose then 10-15 mg/kg every 8 - 12 hours as necessary (maximum 30 mg/kg/day in divided doses),
- **Neonates over 32 weeks corrected gestational age:** 20 mg/kg as a single dose then 10-15 mg/kg every 6 - 8 hours as necessary (maximum 60 mg/kg/day in divided doses),

- **Child 1 month – 6 years:** 20-30 mg/kg as a single dose then 15-20 mg/kg every 4-6 hours as necessary (maximum 75 mg/kg/day in divided doses),
- **Child 6-12 years:** 20-30 mg/kg (max 1 g) as a single dose then 15-20 mg/kg every 4-6 hours as necessary (maximum 75 mg/kg/day or 4 g/day in divided doses),
- **Over 12 years:** 15-20mg/kg (maximum 500mg -1 g) every 4-6 hours as necessary (maximum 4 g /day in divided doses).

Rectal:
- **Neonate 28 – 32 weeks corrected gestational age:** 20 mg/kg as single dose then 10-15 mg/kg every 12 hours as necessary (maximum 30 mg/kg/day in divided doses),
- **Neonates over 32 weeks corrected gestational age:** 30mg/kg as a single dose then 15-20 mg/kg every 8 hours as necessary (maximum 60 mg/kg/day in divided doses),
- **Child 1 – 3 months:** 30 mg/kg as a single dose, then 15-20 mg/kg every 4-6 hours as necessary (maximum 75 mg/kg/day in divided doses),
- **Child 3 months to 12 years:** 30 mg/kg as a single dose (maximum 1g) then 15-20 mg/kg every 4-6 hours as necessary (maximum 75 mg/kg/day or 4 g/day in divided doses),
- **Over 12 years:** 15-20mg/kg (maximum 500mg -1 g) every 4-6 hours as necessary (maximum 4 g/day in divided doses).

IV: as infusion over 15 minutes
- **Preterm neonate over 32 weeks corrected gestational age:** 7.5 mg/kg every 8 hours, maximum 25 mg/kg/day,
- **Neonate:** 10 mg/kg every 4-6 hours (maximum 30 mg/kg/day),
- **Infant and child bodyweight <10kg:** 10mg/kg every 4-6 hours (maximum 30mg/kg/day),
- **Child bodyweight 10-50 kg:** 15 mg/kg every 4-6 hours (maximum 60 mg/kg/day),
- **Bodyweight over 50 kg:** 1g every 4-6 hours (maximum 4 g/day).

Notes:
- Not licensed for use in children under 2 months by mouth; not licensed for use in preterm neonates by intravenous infusion; not licensed for use in children under 3 months by rectum; doses for severe symptoms not licensed; paracetamol oral suspension 500 mg/5 mL not licensed for use in children under 16 years.
- Oral and licensed rectal preparations are licensed for use in infants from 2 months for post immunisation pyrexia (single dose of 60mg which may be repeated once after 4-6hours if necessary), and from 3 months as antipyretic and analgesic.
- IV paracetamol is licensed for short term treatment of moderate pain, and of fever when other routes not possible.
- Consider use of non pharmacological measures to relieve pain, as alternative or in addition to analgesics.
- Hepatotoxic in overdose or prolonged high doses.
- In moderate renal impairment use maximum frequency of 6 hourly; in severe renal impairment maximum frequency 8 hourly.
- Many children and young people with life limiting illness have low weight for their age. The doses above are therefore quoted mainly by weight rather than age (unlike most of the entries in the BNF and BNFc), in order to minimise risk of over-dosing in this patient group.
- Onset of action 15-30 minutes orally, 5-10 minutes IV (analgesia), 30 minutes IV (antipyretic). Duration of action 4-6 hours orally and IV. Oral bioavailability 60-90%. Rectal bioavailability about 2/3 of oral. However, rectal absorption is now known to be erratic and incomplete, and results in slower absorption than oral administration, (except in babies when the oral preparation used rectally speeds absorption compared with suppositories). Elimination is slower in babies under 3 months.
- Dispersible tablets have high sodium content (over 14 mmol per tablet), so caution with regular dosing (consider using the liquid preparation instead).
- For administration via an enteral feeding tube: Use tablets dispersed in water for intragastric or intrajejunal administration. If the sodium content is problematic, use the liquid formulation.

This can be used undiluted for intragastric administration; however, the viscosity of the paediatric liquid preparations is very high; it is difficult to administer these suspensions via a fine bore tube without dilution. If administering intrajejunally, dilute with at least an equal quantity of water to reduce osmolarity and viscosity.

- For management of feverish illness in children, see updated NICE clinical Guideline CG160. (Consider using *either* paracetamol or ibuprofen in children with fever who appear *distressed*, and consider changing to the other agent if distress is not alleviated. But do not use antipyretic agents with the sole aim of reducing body temperature). However, a recent Cochrane systematic review states "there is some evidence that both alternating and combined antipyretic therapy may be more effective at reducing temperatures than monotherapy alone". For babies over 3 months, ibuprofen may be preferable to paracetamol, since asthma seems more common in children who experienced early paracetamol exposure.
- Available as: tablets and caplets (500 mg), capsules (500 mg), soluble tablets (120 mg, 500 mg), oral suspension (120 mg/5 mL, 250 mg/5 mL), suppositories (60 mg, 125 mg, 250 mg, 500 mg and other strengths available from 'specials' manufacturers or specialist importing companies) and intravenous infusion (10 mg/mL in 50mL and 100mL vials).

Evidence: [1-3, 6, 12, 301-305] SR

Paraldehyde (rectal)

Use:
- Treatment of prolonged seizures and status epilepticus.

Dose and route:
By rectal administration (**dose shown is for premixed enema 50:50 with olive oil**)
- **Neonate:** 0.8 mL/kg as a single dose
- **1 month -18 years:** 0.8 mL/kg (maximum 20mL) as a single dose.

Notes:
- Rectal administration may cause irritation.
- Contra-indicated in gastric disorders and in colitis.
- Paraldehyde enema for rectal use is an unlicensed formulation and route of administration.
- Available as paraldehyde enema: premixed solution of paraldehyde in olive oil in equal volumes from 'special-order' manufacturers or specialist importing companies.

Evidence: [2, 6, 306-312] CC, SR

Phenobarbital

Use:
- Adjuvant in pain of cerebral irritation.
- Control of terminal seizures.
- Sedation.
- Epilepsy including status epilepticus. Commonly used first line for seizures in neonates (phenytoin or benzodiazepine are the main alternatives).
- Agitation refractory to midazolam in end of life care.

Dose and routes

Status epilepticus / terminal seizures / agitation
Loading dose: Oral, intravenous or subcutaneous injection:
All ages: 20 mg/kg/dose (maximum 1g) administered over 20 minutes if by IV or SC injection (but see notes below)

Subcutaneous or intravenous injection or infusion:
- **Neonates for control of ongoing seizures:** 2.5-5 mg/kg once or twice daily as maintenance,
- **Child 1 month- 12years:** 2.5-5 mg/kg (maximum single dose 300 mg) once or twice daily or may be given as a continuous infusion over 24 hours,
- **Child 12-18 years:** 300 mg twice daily or may be given as a continuous infusion over 24 hours.

Epilepsy:
By mouth:
- **Neonates for control of ongoing seizures:** 2.5-5 mg/kg once or twice daily as maintenance,
- **Child 1 month–12 years:** 1–1.5 mg/kg twice a day, increased by 2 mg/kg daily as required (usual maintenance dose 2.5–4 mg/kg once or twice a day),
- **Child 12–18 years:** 60–180 mg once a day.

Notes:
- Not licensed for agitation in end of life care.
- Single loading dose is required for initiation of therapy; administer via enteral route if possible. Loading dose can be administered intravenously over 20 minutes or as a slow subcutaneous loading dose however the volume of resultant solution will limit the rate at which a subcutaneous bolus can be administered.
- Loading dose essential to reach steady state quickly and avoid late toxicity due to accumulation.
- For patients already on oral phenobarbital but needing parenteral treatment, doses equivalent to the patient's usual total daily dose of oral phenobarbital can be used.

- Elimination half life of 2 - 6 days in adults, 1 - 3 days in children.
- Phenobarbital induces various enzymes of the CYP450 system and thus may reduce the plasma concentrations of concomitant drugs that are metabolised by this system
- Tablets may be crushed for administration if preferred.
- Use a separate site to commence subcutaneous infusion. SC bolus injections should be avoided because they can cause tissue necrosis due to the high pH.
- It is essential to dilute the injection in 10 times the volume of water for injection before intravenous or subcutaneous injection (i.e. to concentration of 20 mg/mL).
- Available as: tablets (15 mg, 30 mg, 60 mg), oral elixir (15 mg/5 mL) and injection (15mg/mL, 60mg/mL and 200 mg/mL). The licensed oral elixir of 15mg in 5mL contains alcohol 38% and it is preferable to obtain an alcohol free oral liquid via one of the specials manufacturers.

Evidence: [2, 3, 81, 313, 314]

Phenytoin

Use:
- Epilepsy (3rd or 4th line oral antiepileptic) including for status epilepticus.
- Neuropathic pain (effective, at least short term, but not used first line).

Dose
All forms of epilepsy (including tonic-clonic, focal and neonatal seizures) except absence seizures. Neuropathic pain.

Oral or slow IV injection:
- **Neonate:** Initial loading dose by slow IV injection 18 mg/kg **THEN by mouth** 2.5-5 mg/kg twice daily adjusted according to response and plasma phenytoin levels. Usual maximum 7.5 mg/kg twice daily,

- **1 month to 12 years:** initial dose of 1.5-2.5 mg/kg twice daily then adjust according to response and plasma phenytoin levels to 2.5-5 mg/kg twice daily as a usual target maintenance dose. Usual maximum dose of 7.5 mg/kg twice daily or 300 mg daily,
- **12 to 18 years:** initial dose of 75-150 mg twice daily then adjusted according to response and plasma phenytoin levels to 150-200 mg twice daily as a usual target maintenance dose. Usual maximum dose of 300 mg twice daily.

Status epilepticus, acute symptomatic seizures:
Slow IV injection or infusion:
- **Neonate:** 20 mg/kg loading dose over at least 20 minutes, then 2mg/kg/dose (over 30 minutes) every 8-12 hours as a usual maintenance dose in first week of life. Adjust according to response and older babies may need higher doses. After the first dose, oral doses usually as effective as intravenous in babies over 2 weeks old.
- **1 month to 12 years:** 20 mg/kg loading dose over at least 20 minutes, then 2.5-5 mg/kg twice daily usual maintenance dose,
- **12 to 18 years:** 20 mg/kg loading dose over at least 20 minutes, then up to 100mg (over 30 minutes) 3 to 4 times daily usual maintenance dose.

Notes:
- Licensed status: suspension 90 mg in 5 mL is a 'special' and unlicensed. Other preparations are licensed for use in children as an anticonvulsant (age range not specified).
- Phenytoin acts as a membrane stabiliser.
- It has a narrow therapeutic index, unpredictable half life, and the relationship between dose and plasma-drug concentration is non-linear. The rate of elimination is also very variable, especially in the first few weeks and months of life. Co-treatment with commonly used drugs can significantly alter the half life.
- Phenytoin has numerous interactions with other drugs due to hepatic enzyme induction. Long term use is associated with significant side effects. It is no more effective than other anti-epileptics and hence not usually used first line, although it does enable rapid titration.

- Continuous ECG and BP monitoring required during IV administration.
- Oral bioavailability 90-95% is roughly equivalent to intravenous, plasma half-life 7-42 hours. Poor rectal absorption.
- Absorption is exceptionally poor via the jejunal route.
- Reduce dose in hepatic impairment. Monitor carefully if reduced albumin or protein binding e.g. in renal failure.
- Caution: cross-sensitivity is reported with carbamazepine.
- Avoid abrupt withdrawal.
- Consider vitamin D supplementation in patients who are immobilised for long periods or who have inadequate sun exposure or dietary intake of calcium.
- Before and after administration, flush intravenous line with Sodium Chloride 0.9%.
- For *intravenous injection*, give into a large vein at a rate not exceeding 1 mg/kg/minute (max. 50 mg/minute).
- For *intravenous infusion*, dilute to a concentration not exceeding 10 mg/mL with Sodium Chloride 0.9% and give into a large vein through an in-line filter (0.22–0.50 micron) at a rate not exceeding 1 mg/kg/minute (max. 50 mg/minute); complete administration within 1 hour of preparation.
- Prescriptions for oral preparations should include brand name and be of consistent preparation type, to ensure consistency of drug delivery.
- Preparations containing phenytoin sodium are **not** bioequivalent to those containing phenytoin base (such as *Epanutin Infatabs*® and *Epanutin*® suspension); 100 mg of phenytoin sodium is approximately equivalent in therapeutic effect to 92 mg phenytoin base. The dose is the same for all phenytoin products when initiating therapy, however if switching between these products the difference in phenytoin content may be clinically significant. Care is needed when making changes between formulations and plasma phenytoin concentration monitoring is recommended.
- Bioavailability may be reduced unpredictably by enteral feeds and/or nasogastric tube feeds, so flush with water to enhance absorption, interrupt enteral feeding for at least 1-2 hours before

and after giving phenytoin, and maintain similar timings and regimes from day to day.
- Available as tablets (phenytoin sodium 100 mg, generic), capsules (Epanutin^R phenytoin sodium 25 mg, 50 mg,100 mg, 300 mg), Epanutin ^{R} Infatabs (chewable tablets of phenytoin base 50 mg), oral suspension (Epanutin^R phenytoin base 30 mg/5 mL, and 90 mg/5 mL phenytoin base available as an 'unlicensed special'), and injection (phenytoin sodium 50mg/ml generic)

Evidence: [2, 3, 5, 6, 12, 34, 286, 315-319], SR, CC

Phosphate (rectal enema)

Use:
- Constipation refractive to other treatments.

Dose and routes:
By rectal enema:
- **Child 3–7 years:** 45-65 mL once daily,
- **Child 7-12 years:** 65-100 mL once daily,
- **Child 12–18 years:** 100-128 mL once daily.

Notes
- Maintain good hydration and watch for electrolyte imbalance
- Contraindicated in acute gastro-intestinal conditions (including gastro-intestinal obstruction, inflammatory bowel disease, and conditions associated with increased colonic absorption).
- Use only after specialist advice.
- Available as Phosphate enema BP formula B in 128 mL with standard or long rectal tube (NB alternative Fleet^R Ready to Use enema requires slightly different dosing: 40-60ml for age 3-7y, 60-90 ml for age 7-12 y and 90-118 ml for age 12-18y).

Evidence: [1, 2, 320, 321], CC, SR

Promethazine

Use:
- Sleep disturbance.
- Mild sedation
- Antihistamine.
- Can also be used to treat nausea and vomiting, and vertigo.

Dose and routes (for promethazine hydrochloride)
By mouth:

Symptomatic relief of allergy:
- **Child 2–5 years:** 5 mg twice daily *or* 5–15 mg at night,
- **Child 5–10 years:** 5–10 mg twice daily *or* 10–25 mg at night,
- **Child 10–18 years:** 10–20 mg 2–3 times daily *or* 25 mg at night increased to 25 mg twice daily if necessary.

Sedation (short term use):
- **Child 2–5 years:** 15-20 mg at night,
- **Child 5–10 years:** 20-25 mg at night,
- **Child 10–18 years:** 25-50 mg at night.

Nausea and vomiting (particularly in anticipation of motion sickness)
- **Child 2–5 years:** 5 mg twice daily,
- **Child 5–10 years:** 10 mg twice daily,
- **Child 10–18 years:** 20–25 mg twice daily.

Notes:
- Antimuscarinic phenothiazine antihistamine also with D2 antagonist activity.
- Not for use in under 2 years due to risk of fatal respiratory depression.
- Note drug interactions, particularly causing increased antimuscarinic and sedative effects.
- Caution in epilepsy.
- Can be effective for up to 12 hours. Drowsiness may wear off after a few days of treatment.

- For use by feeding tube: the elixir is slightly viscous so can be mixed with an equal volume of water to reduce viscosity and resistance to flushing. Tablets will disintegrate if shaken in water for 5 minutes.
- Available as: promethazine hydrochloride tablets (10 mg, 25 mg), oral elixir (5 mg/5 mL), and injection (25mg/ml). (Promethazine teoclate tablets also available, 25mg, licensed for nausea, vomiting and labyrinthine disorders. Dosing slightly different).

Evidence: [2, 12, 39], CC, EA

Quinine Sulphate

Use:
- Leg cramps.

Dose and routes
By mouth:
- Not licensed or recommended for children as no experience.
- **Adult dose**: quinine sulphate 200mg at bedtime, increased to 300mg if necessary.

Notes:
- Not licensed for use in children for this condition.
- Moderate evidence indicates it to be more effective than placebo in reducing frequency and intensity of cramp.
- Regulatory agencies consider that, given that alternatives to quinine are available, the risks associated with its use are unacceptably high. Rare but serious side effects include thrombocytopenia and haemolytic-uraemic syndrome. Also very toxic in overdose, and has serious interactions with warfarin and digoxin. Therefore MHRA advises that quinine should only be used if 4 criteria are all met: treatable causes have been ruled out, non pharmacological measures have failed, cramps regularly cause loss of sleep, and they are very painful or frequent. Patients should be monitored for signs of thrombocytopenia in the early stages of treatment.

- If used, treatment should be discontinued after 4 weeks if ineffective, and interrupted every 3 months to re-evaluate benefit.
- Available as: tablets (200 mg quinine sulphate).

Evidence: [1, 322, 323], EA

Ranitidine

Use:
- Gastro-oesophageal reflux.
- Treatment of gastritis, benign gastric and duodenal ulcers.
- Gastro-protection (e.g. with combination NSAID/steroids).
- Other conditions requiring reduction in gastric acid.

Dose and routes
By mouth:
- **Neonate:** 2–3 mg/kg 3 times daily (absorption unreliable),
- **Child 1–6 months:** 1 mg/kg 3 times daily increasing if necessary to maximum 3 mg/kg 3 times daily,
- **Child 6 months–3 years:** 2–4 mg/kg twice a day,
- **Child 3–12 years:** 2–4 mg/kg (maximum single dose 150 mg) twice a day. Dose may be increased up to 5 mg/kg (maximum 300 mg/dose) twice daily in severe gastro-oesophageal reflux disease,
- **Child 12–18 years:** 150 mg twice a day or 300 mg at night. May be increased if necessary in moderate to severe gastro-oesophageal reflux disease to 300 mg twice a day or 150 mg 4 times daily for up to 12 weeks.

By slow intravenous injection, diluted to 2.5mg/ml and given over at least 3 minutes (some adult centres give as subcutaneous injection (unlicensed route)):
- **Neonate:** 0.5–1 mg/kg every 6–8 hours (may need 2mg/kg 8 hourly as variable first pass metabolism affects uptake),
- **Child 1 month–18 years:** 1 mg/kg (max. 50 mg) every 6–8 hours (may be given as an intermittent infusion at a rate of 25 mg/hour).

Notes:
- Oral formulations not licensed for use in children < 3 years; injection not licensed for children under 6 months.
- Use gastric pH to judge best dose in early infancy.
- Proton pump inhibitors (PPIs), H2 antagonists and prokinetics all relieve symptoms of non ulcer dyspepsia and acid reflux, PPIs being the most effective. PPIs and 'double dose' H2 antagonists are effective at preventing NSAID-related endoscopic peptic ulcers. Adding a bedtime dose of H2 antagonist to high dose PPI may improve nocturnal acid reflux, but evidence is poor.
- Time to peak plasma concentration is 2-3hours, half-life 2-3hours, duration of action 8-12hours
- Ranitidine may increase plasma concentration of midazolam.
- May cause rebound hyperacidity at night.
- Via feeding tubes, use effervescent tablets as first choice, unless sodium content is a concern. Use oral liquid as alternative. (Standard tablets do not disperse readily in water).
- Can use IV if needed in severe nausea and vomiting. Some centres use subcutaneous doses BD – QDS.
- Available as: tablets and effervescent tablets (150 mg, 300 mg), oral solution (75 mg/5 mL) and injection (25mg/ml).

Evidence: [1-3, 12, 324-327]

Risperidone

Use:
- Dystonia and dystonic spasms refractory to first and second line treatment.
- Psychotic tendency / crises in Battens disease.
- Has anti-emetic activity (some experience in refractory nausea and vomiting in adults; not evaluated in children).
- Treatment of mania or psychosis under specialist supervision.

Dose and routes
Oral:
- **Child 5 - 12 years (weight 20 - 50 kg):** 250 microgram once daily; increasing, if necessary, in steps of 250 microgram on alternate days to maximum of 750 microgram daily.
- **Child 12 years or over (>50 kg):** 500 microgram once daily; increasing in steps of 500 microgram on alternate days to maximum of 1.5 mg daily.
  - In Juvenile Battens Disease, may need 500microgram daily increasing to 1.5mg TDS during crises with hallucinations: this dose can be reduced or stopped as symptoms settle (episodes usually last 1-6 weeks).

Notes
- Not licensed for use in children for these indications. Risperidone is licensed for the short-term symptomatic treatment (up to 6 weeks) of persistent aggression in conduct disorder in children from the age of 5 years.
- 99% bioavailable. 1-2 hours to peak plasma concentration. Onset of action hours to days in delirium; days to weeks in psychosis. Plasma half life 24hours. Duration of action 12-48hours.
- Caution in epilepsy and cardiovascular disease; extrapyramidal symptoms less frequent than with older antipsychotic medications; withdraw gradually after prolonged use. Risperidone can cause significant weight gain.
- Initial and subsequent doses should be halved in renal or hepatic impairment.
- Tablets disintegrate in water within 5 minutes for easy administration via enteral feeding tubes. The oral liquid is simple to administer via feeding tube.
- Available as: tablets (500microgram, 1 mg, 2 mg, 3 mg, 4 mg, 6 mg), orodispersible tablets (500microgram, 1 mg, 2 mg, 3 mg, 4 mg), Liquid 1 mg/mL.

Evidence: CC [2, 12, 124, 328, 329]

Salbutamol

Use:
- Wheezing/ breathlessness caused by bronchospasm.
- Also used in hyperkalemia, for prevention and treatment of chronic lung disease in premature infants, and sometimes in muscular disorders or muscle weakness (seek specialist advice, not covered here).

Dose and routes for exacerbation of reversible airway obstruction, and prevention of allergen- or exercise-induced bronchospasm.

(NB see separate detailed guidance in standard texts for use in acute asthma).

Aerosol Inhalation:
- **Child 1 month-18years:** 100-200 micrograms (1-2 puffs) for persistent symptoms up to four times a day.

Nebulised solution:

- **Neonate:** 1.25-2.5 mg up to four times daily,
- **Child 1 month-18years:** 2.5-5 mg up to four times daily.

Oral preparations: (but use by inhalation preferred for treatment of bronchospasm)
- **Child 1 month–2 years:** 100 micrograms/kg (max. 2 mg) 3–4 times daily
- **Child 2–6 years:** 1–2 mg 3–4 times daily
- **Child 6–12 years:** 2 mg 3–4 times daily
- **Child 12–18 years:** 2–4 mg 3–4 times daily

Notes
- Salbutamol is not licensed for use in hyperkalaemia; syrup and tablets are not licensed for use in children less than 2 years; modified-release tablets are not licensed for use in children less than 3 years; injection is not licensed for use in children.

- In palliative care, if airflow obstruction is suspected, a pragmatic approach may be to give a trial (e.g. 1 – 2 weeks) of a bronchodilator and evaluate the impact on symptoms. Spirometry should normally be used to confirm a possible underlying asthma diagnosis.
- Salbutamol has not been shown to be effective in children less than 2 years, presumably due to the immaturity of the receptors; ipratropium may be more helpful in those less than 1-2 years.
- For an acute episode, many paediatricians now advise multi-dosing of salbutamol 100 microgram up to 10 times, via a spacer where practicable for the patient instead of a nebuliser.
- Side effects: increased heart rate; feeling "edgy" or agitated; tremor.
- The side effects listed above may prevent use, in which case ipratropium bromide is a good alternative.
- Inhaled product should be used with a suitable spacer device, and the child/ carer should be given appropriate training. Inhaler technique should be explained and checked. The HFA (hydrofluoroalkane) propellant now used in multi-dose inhalers tends to clog the nozzle, so weekly cleaning is recommended.
- Salbutamol nebules are intended to be used undiluted. However, if prolonged delivery time (more than 10 minutes) is required, the solution may be diluted with sterile 0.9% NaCl. Salbutamol can be mixed with nebulised solution of ipratropium bromide.
- Available as nebuliser solution (2.5 mg in 2.5 mL, 5 mg in 2.5 mL), respirator solution (5 mg in 1 mL), aerosol inhalation (100 micrograms/puff) by metered dose inhaler (MDI), with various spacer devices. Various types of dry powder inhaler are also available. Also available as salbutamol tablets (costly) 2 mg and 4mg and modified release capsules 4mg and 8mg, and as oral solution 2mg/5ml . Preparations for injection (500micrograms/ml) and intravenous infusion (1mg/ml) are also available.

Evidence: [1-3, 330, 331]

Senna

Use:
- Constipation

Dose and routes
By mouth:
Initial doses which can be adjusted according to response and tolerance
- **Child 1 month –2 years:** 0.5 mL/kg (maximum 2.5 mL) of syrup once a day,
- **Child 2 –6 years:** 2.5-5 mL of syrup a day,
- **Child 6–12 years:** 5-10 mL a day of syrup or 1-2 tablets at night or 2.5-5 mL of granules,
- **Child 12–18 years:** 10-20 mL a day of syrup or 2-4 tablets at night or 5-10 mL of granules.

Notes:
- Syrup is not licensed for use in children < 2 years and tablets/granules are not licensed for use in children <6 years.
- Stimulant laxative.
- Onset of action 8-12 hours.
- Initial dose should be low then increased if necessary.
- Doses can be exceeded on specialist advice.
- Granules can be mixed in hot milk or sprinkled on food.
- Oral liquid may be administered via an enteral feeding tube.
- Available as: tablets (7.5 mg sennoside B), oral syrup (7.5 mg/5 mL sennoside B) and granules (15 mg/5 mL sennoside B).

Evidence: [2, 6, 12, 87]

Sodium Citrate

Use:
- Constipation where osmotic laxative indicated.

Dose and routes

Micolette Micro-enema

Enema, sodium citrate 450 mg, sodium lauryl sulfoacetate 45 mg, glycerol 625 mg, together with citric acid, potassium sorbate, and sorbitol in a viscous solution, in 5-ml
- By rectum: Child 3–18 years: 5–10 mL as a single dose

Micralax Micro-enema

Enema, sodium citrate 450 mg, sodium alkylsulfoacetate 45 mg, sorbic acid 5 mg, together with glycerol and sorbitol in a viscous solution in 5-ml
- By rectum: Child 3–18 years: 5 mL as a single dose

Relaxit Micro-enema

Enema, sodium citrate 450 mg, sodium lauryl sulfate 75 mg, sorbic acid 5 mg, together with glycerol and sorbitol in a viscous solution in a 5ml single dose pack with nozzle.
- By rectum: Child 1 month–18 years: 5 mL as a single dose (insert only half nozzle length in child under 3 years).

Notes
- For under 3 years, insert only half nozzle length.
- Available as: micro-enema (5 mL).

Evidence: [1, 2]

Sodium Picosulfate

Use:
- Constipation.

Dose and routes:
By mouth:
- **Child 1month–4 years:** initial dose of 2.5 mg once a day increasing if necessary according to response to a suggested maximum of 10 mg daily,
- **Child 4–18 years:** initial dose of 2.5 mg once a day increasing if necessary according to response to a suggested maximum of 20 mg daily.

Notes
- Elixir is licensed for use in children of all ages; capsules are not licensed for use in children less than 4 years of age.
- Acts as a stimulant laxative.
- Onset of action 6-12 hours.
- Effectiveness dependent upon breakdown by gut flora – previous effectiveness may therefore be lost during courses of antibiotics and ensuing altered gut flora.
- For administration via an enteral feeding tube: use the liquid preparation; dilute with an equal volume of water prior to administration. Sodium picosulfate reaches the colon without any significant absorption; therefore, the therapeutic response will be unaffected by jejunal administration.
- Available as: elixir (5 mg/5 mL) and capsules (2.5 mg).

Evidence: [1, 2, 12]

Sucralfate

Use:
- Stress ulcer prophylaxis.
- Prophylaxis against bleeding from oesophageal or gastric varices; adjunct in the treatment of: oesophagitis with evidence of mucosal ulceration, gastric or duodenal ulceration, upper GI bleeding of unknown cause.

Dose and route:
Oral
Stress ulcer prophylaxis, prophylaxis against bleeding from oesophageal or gastric varices
- **Child 1 month 2 years:** 250 mg four to six times daily,
- **Child 2-12 years:** 500 mg four to six times daily,
- **Child 12-15 years:** 1 g four to six times daily,
- **Child 15-18 years:** 1 g six times daily (maximum 8 g/day).

Oesophagitis with evidence of mucosal ulceration, gastric or duodenal ulceration
- **Child 1 month -2 years:** 250 mg four to six times daily,
- **Child 2-12 years:** 500 mg four to six times daily,
- **Child 12-15 years:** 1 g four to six times daily,
- **Child 15-18 years:** 2 g twice daily (on rising and at bedtime) or 1 g four times daily (1 hour before meals and at bedtime) taken for 4-6 weeks (up to 12 weeks in resistant cases); maximum 8 g daily.

Notes:
- Not licensed for use in children less than 15 years; tablets are not licensed for prophylaxis of stress ulceration
- Administer 1 hour before meals.
- Spread doses evenly throughout waking hours.
- *Bezoar formation:* Following reports of bezoar formation associated with sucralfate, the CSM has advised caution in seriously ill patients, especially those receiving concomitant enteral feeds or those with predisposing conditions such as delayed gastric emptying.

- Caution – absorption of aluminium from sucralfate may be significant in patients on dialysis or with renal impairment.
- Tablets may be crushed and dispersed in water.
- Administration of sucralfate suspension and enteral feeds via a NG or gastrostomy tube should be separated by **at least** 1 hour. In rare cases bezoar formation has been reported when sucralfate suspension and enteral feeds have been given too closely together.
- Caution – sucralfate oral suspension may block fine-bore feeding tubes.
- Available as: oral suspension (1 g in 5 mL), tablets (1 g).

Evidence: [2, 6, 12]

Temazepam

Use:
- Sleep disturbance where anxiety is a cause.

Dose and routes
By mouth,
- **Adult:** 10–20 mg at night. Dose may be increased to 40 mg at night in exceptional circumstances

Notes:
- Not licensed for use in children.
- Oral solution may be administered via an enteral feeding tube.
- Available as: tablets (10 mg, 20 mg) and oral solution (10 mg/5 mL).
- Schedule 3 controlled drug

Evidence: [1, 12]

Tizanidine

Use:
- Skeletal muscle relaxant.
- Chronic severe muscle spasm or spasticity.

Dose and routes
Children doses based on SR [332]
- **Child 18 months – 7 years:** 1 mg/day; increase if necessary according to response,
- **Child 7 -12 years:** 2 mg/day; increase if necessary according to response,
- **Child >12 years:** as per adult dose [1]:Initially 2 mg increasing in increments of 2 mg at intervals of 3–4 days. Give total daily dose in divided doses up to 3–4 times daily. Usual total daily dose 24 mg. Maximum total daily dose 36 mg.

Notes:
- Not licensed for use in children.
- Usually prescribed and titrated by neurologists.
- Timing and frequency of dosing is individual to the specific patient as maximal effect is seen after 2–3hours and is short-lived.
- Use with caution in liver disease, monitor liver function regularly.
- Use with caution with drugs known to prolong the QT interval.
- Avoid abrupt withdrawal – risk of rebound hypertension and tachycardia.
- Tizanidine plasma concentrations are increased by CYP1A2 inhibitors potentially leading to severe hypotension.
- Drowsiness, weakness and dry mouth are common side-effects.
- Tablets may be crushed and administered in water if preferred. May be administered via an enteral feeding tube - Tablets do not disperse readily, but will disintegrate if shaken in 10 mL of water for 5 minutes. The resulting dispersion will flush via an 8Fr NG tube without blockage.
- Available as: tablets (2 mg, 4 mg).

Evidence: [1, 12, 19, 20, 25, 332-335]

Tramadol

The WHO now advises there is insufficient evidence to make a recommendation for an alternative to codeine (tramadol) and recommends moving directly from non-opioids (Step 1) to low dose strong opioids for the management of moderate uncontrolled pain in children..

Use:
- Minor opioid with additional non-opioid analgesic actions.

Dose and routes
By mouth:
- **Child 5-12 years:** 1-2 mg/kg every 4-6 hours (maximum initial single dose of 50 mg; maximum of 4 doses in 24 hours). Increase if necessary to a maximum dose of 3 mg/kg (maximum single dose 100 mg) every 6 hours,
- **Child 12–18 years:** initial dose of 50 mg every 4–6 hours. Increase if necessary to a maximum of 400 mg/day given in divided doses every 4-6 hours.

By IV injection or infusion
- **Child 5-12 years:** 1-2 mg/kg every 4-6 hours (maximum initial single dose of 50 mg; maximum 4 doses in 24 hours). Increase if necessary to a maximum dose of 3 mg/kg (maximum single dose 100mg) every 6 hours,
- **Child 12-18 years:** initial dose of 50 mg every 4-6 hours. Dose may be increased if necessary to 100 mg every 4-6 hours. Maximum 600 mg/DAY in divided doses.

Notes:
- Not licensed for use in children < 12 years.
- Tramadol is a Schedule 3 CD but exempted from safe custody requirements.
- By mouth tramadol is about 1/10 as potent as morphine.
- Onset of action after an oral dose is 30 to 60 minutes. Duration of action is 4-9 hours.

- Causes less constipation and respiratory depression than the equivalent morphine dose.
- Analgesic effect is reduced by ondansetron.
- Soluble or orodispersible tablets may be dissolved in water for administration via an enteral feeding tube.
- Available as capsules (50 mg, 100 mg), soluble tablets (50 mg), orodispersible tablets (50 mg), m/r tablets and capsules (100 mg, 150 mg, 200 mg, 300 mg, 400 mg), oral drops (100mg/mL) and injection (50 mg/mL).

Evidence: [1, 2, 12, 29, 33]

Tranexamic acid

Use:
- Oozing of blood (e.g. from mucous membranes / capillaries), particularly when due to low or dysfunctional platelets.
- Menorrhagia.

Dose and routes

By mouth:
Inhibition of fibrinolysis
- **Child 1 month–18 years:** 15–25 mg/kg (maximum 1.5 g) 2–3 times daily.

Menorrhagia
- **Child 12-18 years:** 1 g 3 times daily for up to 4 days. If very heavy bleeding a maximum daily dose of 4 g (in divided doses) may be used. Treatment should not be initiated until menstruation has started.

By intravenous injection over at least 10 minutes:

Inhibition of fibrinolysis
- **Child 1month -18 years:** 10 mg/kg (maximum 1 g) 2-3 times a day.

By continuous intravenous infusion:

Inhibition of fibrinolysis
- **Child 1month -18 years:** 45 mg/kg over 24 hours.

Mouthwash 5% solution:
- **Child 6-18 years:** 5-10 mL 4 times a day for 2 days. Not to be swallowed.

Topical treatment:
- Apply gauze soaked in 100mg/mL injection solution to affected area.

Notes:
- Injection not licensed for use in children under 1 year or for administration by intravenous infusion.
- Can cause clot 'colic' if used in presence of haematuria.
- Parenteral preparation can be used topically.
- Available as: tablets (500 mg), syrup (500 mg/5mL available from 'specials' manufacturers) and injection (100 mg/mL 5 mL ampoules). Mouthwash only as extemporaneous preparation.

Evidence: [2, 6, 336-340]

Trihexyphenidyl

Uses:
- Dystonias; Sialorrhoea (drooling); Antispasmodic.

Dose and route:
Oral
- **Child 3 months -18 years:** initial dose of 1-2mg daily in 1-2 divided doses, increased every 3-7 days by 1mg daily; adjusted according to response and side-effects; maximum 2mg/kg/daily (maximum 70mg/daily).

Generally, the doses needed to control drooling are much lower than those needed for dystonias.

Notes:
- Anticholinergic agent thought to act through partially blocking central (striatal) cholinergic receptors.
- Not licensed for use in children.
- Use in conjunction with careful observation and a full non-drug management programme including positioning, massage, holding, distraction, checking for causes of exacerbations etc. Advisable to seek specialist neurological input before use of trihexyphenidyl.
- Side-effects are very common and it is important to start at a low dose and increase gradually to minimise the incidence and severity. Mouth dryness, GI disturbance, blurring of vision, dizziness and nausea can occur in 30-50% patients. Less common side-effects include urinary retention, tachycardia and with very high doses CNS disturbance.
- Use with caution in children with renal or hepatic impairment.
- Onset of action is usually within 1 hour, maximum effect occurs within 2-3 hours and duration of effect ~6-12 hours.
- May take several weeks for maximal effect on dystonic movements to be seen.
- Do not withdraw abruptly in children who have been on long-term treatment.
- Tablets may be crushed and mixed in soft food.
- For administration via a gastrostomy the liquid may be used or the tablets will disperse readily in water.
- Available as: tablets 2mg and 5mg; oral liquid (pink syrup) 5mg in 5ml.

Reference: [1, 2, 12, 341-347]

Vitamin K (Phytomenadione)

Use:
- Treatment of haemorrhage associated with vitamin-K deficiency (seek specialist advice).

Dose and routes
By mouth or intravenous:
- **Neonate**: 100 micrograms/kg.
- **Child 1 month–18 year**: 250-300 micrograms/kg (maximum 10 mg) as a single dose.

Notes:
- Caution with intravenous use in premature infants <2.5 kg.
- Available as Konakion MM injection 10 mg/mL (1 mL amp) for slow intravenous injection or intravenous infusion in glucose 5%; NOT for intramuscular injection.
- Available as Konakion MM Paediatric 10 mg/mL (0.2 mL amp) for oral administration or intramuscular injection. Also for slow intravenous injection or intravenous infusion in glucose 5%.
- There is not a UK licensed formulation of Vitamin K tablets currently available. Possible to obtain 10 mg phytomenadione tablets via a specialist importation company.

Evidence:[1-3, 6]

Appendices

Appendix 1: Morphine equivalence single dose [1, 2, 5]

Analgesic	Dose
Morphine oral	10mg
Morphine subcutaneous	5mg
Diamorphine subcutaneous	3mg
Hydromorphone oral	2mg
Oxycodone oral	6.7mg
Methadone	Variable

Appendix 2: Subcutaneous infusion drug compatibility

Evidence suggests that in during end of life care in children, where the enteral route is no longer available, the majority of symptoms can be controlled by a combination of six "essential drugs" [216]Compatibility for these six drugs is given in the table 1 below [8]. For more detailed information professionals are advised to consult an appropriate reference source [217]

Table 1: Syringe driver compatibility for two drugs in water for injection

diamorphine

-	morphine sulphate					
+	+	midazolam				
A+	+	+	cyclizine			
A+	+	+	+	haloperidol		
+	?	+	-	-	levomepromazine	
+	+	+	?	?	?	hyoscine hydrobromide

A	Laboratory data; physically and chemically compatible but crystallization may occur as concentrations of either drug increase
+	Compatible in water for injection at all usual concentrations
-	Combination not recommended; drugs of similar class
?	No data available

Appendix 3: Template Symptom Management Plan

NAME:
DATE OF BIRTH:
ID NUMBER:
WEIGHT:
DRUG ALLERGIES:
DOCTOR/NURSE COMPLETING THE FORM:

DIAGNOSIS:

BEST, MOST LIKELY AND WORST CASE SCENARIOS:

MAIN PROBLEM LIST:

Pain Management Plan Template

Possible Causes of Pain:

Children may suffer with pain due to local tissue damage, inflammation, muscle spasm, nerve irritation or damage, organ distension or the build-up of pressure within a closed space (such as the skull or liver capsule), secondaries in bone or elsewhere, muscle spasm, colic /constipation, retention of urine or psychological reasons

Non-pharmacological Management

Not all pain can or needs to be treated with drugs. Children are uniquely open to non-pharmacological approaches such as distraction, hypnosis, progressive muscular relaxation, singing, praying, and (of course) plenty of cuddling. Complementary medical treatment such as massage, reflexology, acupuncture and herbal treatments may also be helpful. There are no rules in CPC – we are inherently practical, so we do whatever works help the child.

Pharmacological Management

We work according to the WHO pain ladder:
1. Step 1: Paracetamol and Ibuprofen or Diclofenac
2. Step 2: Strong opioids such as morphine, diamorphine, oxycodone, fentanyl and buprenorphine
3. All along: Adjuvants such as amitriptyline or anticonvulsants for nerve pain, dexamethasone for pain due to pressure or compression, hyoscine for smooth muscle pain, and some specific drugs such as ketamine or nitrous oxide.

SPECIFIC PAIN MANAGEMENT PLAN FOR THIS CHILD

Current problems for this child

Specific management plan for this child (please give correct doses)

Step 1:

Step 2:

Step 3:

If the child continues to suffer with these problems despite these steps, please call xxx

Nausea and Vomiting Management Plan Template

Possible Causes of Nausea and Vomiting

Children may become nauseated due to anxiety, local problems with the stomach and intestine, certain drugs, general systemic illness, chemical imbalances, damage to the vestibular system, raised or raised intracranial pressure.

Non-pharmacological Management

- As with pain, nausea and vomiting does not always need to be treated with drugs.
- The same non-pharmacological approaches apply; such as distraction, hypnosis, progressive muscular relaxation, singing, praying, and (of course) plenty of cuddling.
- Complementary medical treatment such as massage, reflexology, acupuncture and herbal treatments may also be helpful.
- Other practical steps include removing unpleasant smells and food that is finished, avoiding strong perfumes and using small amounts of food more frequently.

Pharmacological Management

- Vomiting is triggered by different pathways involving the gut, the inner ear, the vomiting centre in the brain and the higher cortical centres.
- Each of these uses different receptors and chemical transmitters, and so each requires a different drug: cyclizine for higher and vestibular causes, metoclopramide and domperidone for upper gastric causes, haloperidol and ondansatron for chemical causes, and levomepromazine for all causes (although this drug is very sedating).

SPECIFIC NAUSEA AND VOMITING MANAGEMENT PLAN FOR THIS CHILD

Current problems for this child

Specific management plan for this child (please give correct doses)

Step 1:

Step 2:

Step 3:

If the child continues to suffer with these problems despite these steps, please call xxx

Seizure Management Plan Template

Possible Causes of seizures

- Children may have seizures as a result of their underlying condition, high fevers, brain irritation or infection, and raised intracranial pressure. They often stop without intervention, but should be treated if they last beyond 3-4 minutes (which seems like a very long time when a child is fitting).
- In the CPC setting we see children in one of two categories.
  - Children with global neurological disorders who have had seizures for a long time, on many drugs, and whose families feel able and equipped to manage them
  - Children with no history of seizures where there illness has started to trigger seizures. These are particularly frightening for the families, and also care staff

Non Pharmacological Management

- The most useful thing is calm, clear and gentle education and training for families so they know what to expect and what to do. A written plan is very useful for the family and school to have with them.
- If the child does fit, make sure that children are in a safe position and is not going to hurt themselves on anything. Do not try to stop movements and do not try to put anything into their mouths. Give oxygen (if available). Look for reversible causes of increased seizures and try to correct them e.g. infection, biochemical imbalance, hypoglycaemia, raised intracranial pressure and inappropriate epilepsy management. Not all seizures require drugs, so wait 5 minutes (which will seem like an age) to see if the child stops fitting naturally)

Diazepam rectal doses for seizures
- Neonate: 1.25–2.5mg repeated once after 10min if necessary

- Child 1 month–2 years: 5mg repeated once after 10min if necessary
- Child 2–12 years: 5–10mg repeated once after 10min if necessary
- Child 12–18 years: 10mg-20mg repeated once after 10min if necessary

Pharmacological Management

- If a child starts fitting and if you have oxygen, give it at high flow. If the child does not stop fitting give midazolam buccally (or diazepam rectally, or paraldehyde mixed with oil rectally).
- If there is no improvement after another 10 minutes give a second dose and ring the ambulance or palliative care team.

Midazolam buccal doses for seizures
- Neonate: 300microgram/kg as a single dose
- Child 1–6 months: 300microgram/kg (max. 2.5mg), repeated once if necessary
- Child 6 months–1 year: 2.5mg, repeated once if necessary;
- Child 1–5 years: 5mg, repeated once if necessary
- Child 5–10 years: 7.5mg, repeated once if necessary
- Child 10–18 years: 10mg, repeated once if necessary.

SPECIFIC SEIZURE MANAGEMENT PLAN FOR THIS CHILD

Current problems for this child

Specific management plan for this child (please give correct doses)

Step 1:

Step 2:

Step 3:

If the child continues to suffer with these problems despite these steps, please call xxx

Acute Breathlessness Management Plan Template

Possible Causes of Breathlessness

- Children may have breathlessness as a result of their underlying condition, breathlessness is a subjective experience if breathing discomfort that varies in intensity.
- Treatment of any underlying condition is therefore important (e.g. anxiety, bronchospasm, pressure on the diaphragm from ascites, anaemia, pleural effusions etc.)

Non-pharmacological Management

- Circulation of air- using a fan or open window.
- Anxiety can cause the exacerbation of breathlessness therefore explanations and reassurance should be given to parents and the child.
- Positioning - should be kept as upright as possible.
- Age appropriate distraction is important, including play and breathing exercises (for older children).

Pharmacological Management

- **Oxygen**
    - Studies have shown that oxygen therapy may not be beneficial unless the breathlessness is related to an acute desaturation although there may be some benefits for comfort and reassurance.
- **Morphine**
    - This reduces anxiety, pain and pulmonary artery pressure. Begin with half the analgesic dose and titrate to effect (See section on pain).

- **Midazolam**:
    - Buccal: 1 month- 18 years: 200-500 micrograms/kg (max 10mg)
    - Rectal: 500 – 750 mcg/kg
- **Continuous intravenous/ subcutaneous infusion: 1 month-18 years: 10-300 micrograms/kg/h.**
- **Lorazepam**
    - All ages 25-50 micrograms/kg single dose.
    - Most children will not need more than 0.5-1mg for a trial dose. Well absorbed sublingually.

SPECIFIC BREATHLESSNESS MANAGEMENT PLAN FOR THIS CHILD

Current problems for this child

Specific management plan for this child (please give correct doses)

Step 1:

Step 2:

Step 3:

If the child continues to suffer with these problems despite these steps, please call xxx

Excessive Respiratory Secretions Management Plan Template

Possible Causes of Respiratory Secretions

- Excessive respiratory secretions can often cause distress to the child and carers in the terminal phase.

Non-pharmacological Management

- Positioning the child carefully can often reduce noisy breathing considerably.
- Suction may be helpful in a few children, but is only recommended in the oral cavity.
- Explanation and reassurance is beneficial

Pharmacological Management

- Excessive respirator secretions may be reduced by Hyoscine Hydrobromide patches or by subcutaneous injection or infusion. The use of Glycopyrronium bromide may also be considered.

Hyoscine hydrobromide transdermal doses:
1month – 3 years: 250 micrograms every 72 hours (quarter of a patch).
3-10 years: 500 micrograms every 72 hours (half a patch)
10-18 years: 1mg every 72 hours (1 patch)

Hyoscine hydrobromide subcutaneous/ intravenous infusion:
10 mcg/kg every 4-8 hours, maximum 600mcg.
Or 60 mcg/kg over 24h.

Glycopyrronium bromide oral doses:
1 month- 18 years: 40-100 micrograms/kg (max of 2mg). 3-4 times daily.

Glycopyrronium bromide subcutaneous infusion
Child 1 month- 12 years: 12-40micrograms/kg/24 hours (max 1.2mg)
Child 12-18 years: 0.6-1.2mg/ 24 hours.

SPECIFIC RESPIRATORY SECRETIONS MANAGEMENT PLAN FOR THIS CHILD

Current problems for this child

Specific management plan for this child (please give correct doses)

Step 1:

Step 2:

Step 3:

If the child continues to suffer with these problems despite these steps, please call xxx

Severe Bleeding Management Plan Template

Possible Causes of Bleeding:

- Children may suffer from bleeding due to certain drugs, a number of malignancies, liver diseases, clotting disorders, ulcers, renal failure, general systemic infections, bleeding gums.

Non Pharmacological Management:

- Heavy bleeding is very frightening for the child and family. Try not to panic.
- Apply pressure to bleeding wounds.
- Position children so blood can train more easily.
- Use dark coloured sheets and towels to mop blood so as not to cause distress to witnesses.
- Distraction techniques such as singing, breathing or relaxation exercises, playing or spiritual activities such as praying.

Pharmacological Management

- If an acute or severe bleed has occurred emergency fast-acting sedative medication using midazolam and diamorphine used. These should be easily and rapidly available in the child's home.
- As soon as possible, set up a continuous subcutaneous or intravenous infusion of midazolam 0.3mg/kg/24hrs and morphine or diamorphine at a dose that is at least the equivalent of an intravenous breakthrough pain dose. Don't be scared to increase the doses rapidly as required.
- Remember there is no maximum dose of morphine for pain, and events can change rapidly during the end of life stage. Be prepared to anticipate and manage any side effects of morphine (nausea, vomiting, constipation etc.) that may occur.

- External dressings can be applied on their own or soaked in adrenaline or tranexamic acid.
- Internal bleeding can be treated with clotting treatments such as vitamin K. If extensive blood loss a blood transfusion/platelet transfusion may be considered.

Midazolam: IV (or buccal or nasal if no IV access) midazolam 100mcg/kg (children less than five may need much more, up to 600mg/kg but start low and be ready to repeat or titrate up). Repeat every 10 minutes until sedation is complete. For less acute presentations consider diazepam PR.

Diamorphine: IV (or buccal or nasal if no IV access) diamorphine 0.1 mg/kg (halve that for children under six months) and repeat as above.

SPECIFIC BLEEDING MANAGEMENT PLAN FOR THIS CHILD

Current problems for this child

Specific management plan for this child (please give correct doses)

Step 1:

Step 2:

Step 3:

If the child continues to suffer with these problems despite these steps, please call xxx

Muscle Spasm (Dystonia) Management Plan Template

Possible Causes of Dystonia

- Muscle spasms can be caused by pressure on or irritation of a nerve. Spasms can also be caused by a number of life limiting conditions such as neurodegenerative disorders. Prolonged periods or bed rest or inactivity can cause muscle spasms.

Non Pharmacological Management:

- It is also important to be aware of a vicious circle that can arise: dystonia is very painful and frightening, and pain and anxiety causes muscles to tense up, making the dystonia and hence pain and anxiety worse.
- The key to dystonia management is therefore trying to break the cycle.
- Treating any cause of anxiety or pain is likely to help.
- Muscle spasms do not always need to be treated with medication alone, and medication is often not very effective, or causes a lot of side effects.
- Trying to help the child to relax, through positioning, correct seating, massage, physiotherapy, warm bathing, or distraction are at least as important as medication, and possibly more so.

Pharmacological Management

- Skeletal muscle relaxants can be helpful, but tend to have a lot of side effects.
- At the end of life, benzodiazepines such as diazepam or midazolam are probably the most useful and easiest to use, but long term use is more difficult as children can develop tolerance to them. In

such cases baclofen (oral or intrathecal), tizanidine, dantrolene, botulinum injections, and local surgery may be used.

Diazepam:

1m – 2y: start at 0.25mg/kg bd and titrate upwards as needed
2-5y: 2.5 mg bd and titrate upwards as needed
5-12y: 5mg bd and titrate upwards as needed
12-18y: 5-15mg bd and titrate upwards as needed

SPECIFIC DYSTONIA MANAGEMENT PLAN FOR THIS CHILD

Current problems for this child

Specific management plan for this child (please give correct doses)

Step 1:

Step 2:

Step 3:

If the child continues to suffer with these problems despite these steps, please call xxx

Mouth Pain Management Plan Template

Possible Causes of Mouth Pain:

- Mouth care is often overlooked in children's palliative care. The possible causes include oral candidiasis; dry mouth due to mouth breathing, oxygen usage and side effects to medications; traumatic/aphthous/infectious ulcers; bleeding gums; dental caries and gum hyperplasia.

Non-Pharmacological Management

- As with other symptoms, mouth pain does not always need to be treated with medications.
- Keeping the mouth clean and moist can prevent the mouth from becoming dry. This can be achieved by sucking ice, pineapple or an ice lolly, sipping cold water, dipping gauze or pink sponge in water or mouth wash.
- Placing small pieces of chocolate in the mouth if appropriate can be pleasurable.
- Other non-pharmacological approaches such as distraction, hypnosis, progressive muscular relaxation, singing, praying and of course plenty of cuddling.
- Complementary medical treatment such as massage, reflexology, acupuncture and herbal treatments may also be helpful.

Pharmacological Management

To alleviate mouth pain it is necessary to identify and treat the cause. Please also refer to the pain management symptom control plan if necessary.

Dry mouth	Sips of fluids, lollies, artificial saliva
Oral candidiasis	Miconazole oral gel or fluconazole
Mouth ulcer	Mouthwash, bonjela, steroid lozenges
Bleeding gums	Tranexamic Acid, mouth washes, or haemostatic agents. e.g. gel foam / gel film. Refer to bleeding management plan

SPECIFIC MANAGEMENT PLAN FOR THIS CHILD

Current problems for this child

Specific management plan for this child (please give correct doses)

Step 1:

Step 2:

Step 3:

If the child continues to suffer with these problems despite these steps, please call xxx

Spinal Cord Compression Management Plan Template

Possible Causes of Spinal Cord Compression:

- Spinal Cord Compression is a medical emergency. The aim of treatment is to prevent loss of mobility and/or continence. This may not be a relevant aim in all patients, so the treatment choice depends on the answers to 2 questions:
    1. Is spinal cord compression possible? Be alert to any child with spinal disease who develops worsening pain, neurological symptoms in the legs or bowel/bladder changes.
    2. Is it in the child's best interests to carry out emergency investigation (e.g. MRI scan) and treatment (e.g. surgery or radiotherapy)
- The most likely cause of spinal cord compression is spinal metastases. The signs of spinal cord compression include back pain, weak legs, altered reflexes, weakness sensation in legs, bladder and anal sphincter disturbance.

Non-Pharmacological Management

- Mobility management
- Skin care
- Urinary system management
- Psychosocial support
- Other non-pharmacological approaches such as distraction, hypnosis, progressive muscular relaxation, singing, praying and of course plenty of cuddling.
- Complementary medical treatment such as massage, reflexology, acupuncture and herbal treatments may also be helpful.

Pharmacological Management

- Pain should be treated (see the relevant section). If it is in the child's best interests to have further investigation and treatment.
- Dexamethasone IV should be administered and the child admitted immediately.

Dexamethasone
Under 35 kg: 6-8mg
Over 35kg: 12-16mg

SPECIFIC MANAGEMENT PLAN FOR THIS CHILD

Current problems for this child

Specific management plan for this child (please give correct doses)

Step 1:

Step 2:

Step 3:

If the child continues to suffer with these problems despite these steps, please call xxx

Agitation, Restlessness, Delirium and Anxiety Management Plan Template

Possible Causes of Agitation and Anxiety

- There are many possible causes of agitation, restlessness, anxiety and delirium in palliative care.
- These include urinary retention, constipation, gastro-oesophageal reflux, seizures, muscle spasm, medications, unidentified injury/fracture, sepsis, cerebral causes, hypoxia, environment, fear, nausea, uncontrolled pain and poor positioning.

Non Pharmacological Management

Treat the underlying cause if present. Try one or more of the following:
- Reassurance
- Reposition
- Involve parents
- Keep safe
- Adjust environment as necessary
- Relaxing music

Pharmacological Management

Treat specific cause if known and also treat severe cases by giving:
- Buccal Midazolam: 200mcg – 500mcg/kg
- Diazepam:
    - Under 4y: 50mcg/kg bd
- Haloperidol: 12.5-25mcgs/kg (max 10mg/24h)
- Levomepromazine:
    - 2-12y: 0.25-1mg/kg nocte
    - >12y: 6.25-25mg nocte
    - 5-12y: 1.5-10mg bd
    - >12: 2-10mg bd
- Levomepromazine: (more sedating but can also be given on a regular basis)

SPECIFIC MANAGEMENT PLAN FOR THIS CHILD

Current problems for this child

Specific management plan for this child (please give correct doses)

Step 1:

Step 2:

Step 3:

If the child continues to suffer with these problems despite these steps, please call xxx

Constipation Management Plan Template

Possible Causes of Constipation

The definition of constipation in children in palliative care can be considered to be a change in the normal bowel habits of a child. Oral preparations are normally preferable to rectal medication. It is important to bear in mind that a child may not tolerate any form of rectal examination.

- Inactivity
- Dehydration (fever/vomiting)
- Dietary considerations
- Disease/condition specific – neurological/metabolic
- Bowel obstruction/rectal tears
- Medication

Non Pharmacological Management

- Extra fluids (e.g. water, prune juice, fibre rich milk)
- Massage
- Position change
- Comforting/distraction
- Warm bag

Pharmacological Management

- Bisacodyl suppository or enema (first line of treatment to relieve severe constipation)
- Senna (stimulant laxative for a short period)
- Movicol and/or Lactulose (osmotic laxative in order to prevent further episodes of constipation)

SPECIFIC MANAGEMENT PLAN FOR THIS CHILD

Current problems for this child

Specific management plan for this child (please give correct doses)

Step 1:

Step 2:

Step 3:

If the child continues to suffer with these problems despite these steps, please call xxx

Formulary References

1. BNF, *British National Formulary*, ed. R. BMA. 2013, London: BMJ Publishing Group, RPS Publishing,.
2. BNF, *British National Formulary for Children*, ed. R. BMA, RCPCH, NPPG. 2013-14, London: BMJ Publishing Group, RPS Publishing, and RCPCH Publications.
3. NNF6, *Neonatal Formulary 6*. BMJ Books, ed. E. Hey. 2011: Blackwell Publishing.
4. WHO, *WHO guidelines on the pharmacological treatment of persisting pain in children with medical illnesses*. 2012.
5. Twycross R and Wilcock A, *Palliative Care Formulary (PCF 4)*. 4th ed. 2011: Nottingham: Palliativedrugs.com Ltd.
6. RCPCH, N., *'Medicines for Children'*. 2nd ed. ed. 2003: RCPCH Publications limited.
7. Von Heijne, M., et al., *Propofol or propofol--alfentanil anesthesia for painful procedures in the pediatric oncology ward*. Paediatr Anaesth, 2004. **14**(8): p. 670-5.
8. Duncan, A., *The use of fentanyl and alfentanil sprays for episodic pain*. Palliat Med, 2002. **16**(6): p. 550.
9. Selby & York Palliative Care Team & Pharmacy Group. *Prescribing and administration information for Alfentanil spray* 2007; Available from: www.yacpalliativecare.co.uk/documents/download21.pdf
10. Hershey, A.D., et al., *Effectiveness of amitriptyline in the prophylactic management of childhood headaches*. Headache, 2000. **40**(7): p. 539-49.
11. Heiligenstein, E. and B.L. Steif, *Tricyclics for pain*. J Am Acad Child Adolesc Psychiatry, 1989. **28**(5): p. 804-5.
12. Rebecca White and Vicky Bradnam, *Handbook of Drug administration via Enteral Feeding Tubes*. 2nd ed, ed. B.P.N. Group. 2011: Pharmaceutical Press.
13. Gore, L., et al., *Aprepitant in adolescent patients for prevention of chemotherapy-induced nausea and vomiting: a randomized, double-blind, placebo-controlled study of efficacy and tolerability*. Pediatr Blood Cancer, 2009. **52**(2): p. 242-7.

14. Murphy D et al, *Aprepitant is efficacious and safe in young teenagers.* . Pediatr Blood Cancer, 2011. **57**(5): p. 734-735 (Abs).
15. Williams D et al, *Extended use of aprepitant in pediatric patients.* Biology of Blood and Marrow Transplantation, 2012. **18**(2): p. Suppl 2 S378 (Abs).
16. Choi, M.R., C. Jiles, and N.L. Seibel, *Aprepitant use in children, adolescents, and young adults for the control of chemotherapy-induced nausea and vomiting (CINV).* J Pediatr Hematol Oncol, 2010. **32**(7): p. e268-71.
17. Murphy C et al, *NK1 receptor antagonism ameliorates nausea and emesis in typical and atypical variants of treatment refractory cyclical vomiting syndrome.* J Pediatr Gastroenterology Nutr,, 2006. **42**(5): p. e13-14.
18. Dachy, B. and B. Dan, *Electrophysiological assessment of the effect of intrathecal baclofen in dystonic children.* Clin Neurophysiol, 2004. **115**(4): p. 774-8.
19. Campistol, J., *[Orally administered drugs in the treatment of spasticity].* Rev Neurol, 2003. **37**(1): p. 70-4.
20. Delgado, M.R., et al., *Practice parameter: pharmacologic treatment of spasticity in children and adolescents with cerebral palsy (an evidence-based review): report of the Quality Standards Subcommittee of the American Academy of Neurology and the Practice Committee of the Child Neurology Society.* Neurology. **74**(4): p. 336-43.
21. Gormley, M.E., Jr., L.E. Krach, and L. Piccini, *Spasticity management in the child with spastic quadriplegia.* Eur J Neurol, 2001. **8 Suppl 5**: p. 127-35.
22. Hansel, D.E., et al., *Oral baclofen in cerebral palsy: possible seizure potentiation?* Pediatric Neurology, 2003. **29**(3 SU -): p. 203-206.
23. Jones, R.F. and J.W. Lance, *Bacloffen (Lioresal) in the long-term management of spasticity.* Med J Aust, 1976. **1**(18): p. 654-7.
24. Pascual-Pascual, S.I., *[The study and treatment of dystonias in childhood].* Rev Neurol, 2006. **43 Suppl 1**: p. S161-8.
25. Patel, D.R. and O. Soyode, *Pharmacologic interventions for reducing spasticity in cerebral palsy.* Indian J Pediatr, 2005. **72**(10): p. 869-72.
26. Drugs.com, *http://www.drugs.com/dosage/bethanechol.html.* 2014.
27. Durant, P.A. and T.L. Yaksh, *Drug effects on urinary bladder tone during spinal morphine-induced inhibition of the micturition reflex in unanesthetized rats.* Anesthesiology, 1988. **68**(3): p. 325-34.

28. Attina, G., et al., *Transdermal buprenorphine in children with cancer-related pain.* Pediatr Blood Cancer, 2009. **52**(1): p. 125-7.
29. Zernikow, B., et al., *Pediatric palliative care: use of opioids for the management of pain.* Paediatr Drugs, 2009. **11**(2): p. 129-51.
30. Dahan, A., L. Aarts, and T.W. Smith, *Incidence, Reversal, and Prevention of Opioid-induced Respiratory Depression.* Anesthesiology, 2010. **112**(1): p. 226-38.
31. Colvin, L. and M. Fallon, *Challenges in cancer pain management--bone pain.* Eur J Cancer, 2008. **44**(8): p. 1083-90.
32. Kienast, H.W. and L.D. Boshes, *Clinical trials of carbamazepine in suppressing pain.* Headache, 1968. **8**(1): p. 1-5.
33. Klepstad, P., et al., *Pain and pain treatments in European palliative care units. A cross sectional survey from the European Association for Palliative Care Research Network.* Palliat Med, 2005. **19**(6): p. 477-84.
34. Swerdlow, M., *The treatment of "shooting" pain.* Postgrad Med J, 1980. **56**(653): p. 159-61.
35. Lynch, P.M., et al., *The safety and efficacy of celecoxib in children with familial adenomatous polyposis.* Am J Gastroenterol. **105**(6): p. 1437-43.
36. Foeldvari, I., et al., *A prospective study comparing celecoxib with naproxen in children with juvenile rheumatoid arthritis.* J Rheumatol, 2009. **36**(1): p. 174-82.
37. Stempak, D., et al., *Single-dose and steady-state pharmacokinetics of celecoxib in children.* Clin Pharmacol Ther, 2002. **72**(5): p. 490-7.
38. Drugs.com, *http://www.drugs.com/dosage/celecoxib.html.* 2014.
39. Jones, D.P. and E.A. Jones, *Drugs for Insomnia.* Can Med Assoc J, 1963. **89**: p. 1331.
40. Pandolfini, C. and M. Bonati, *A literature review on off-label drug use in children.* Eur J Pediatr, 2005. **164**(9): p. 552-8.
41. Weiss, S., *Sedation of pediatric patients for nuclear medicine procedures.* Semin Nucl Med, 1993. **23**(3): p. 190-8.
42. Friedman, N.L., *Hiccups: a treatment review.* Pharmacotherapy, 1996. **16**(6): p. 986-95.
43. Jassal, S., ed. *Basic Symptom Control in Paediatric Palliative Care.* 9th ed. Rainbow's Hospice Symptom Control Manual, ed. S. Jassal. 2013.
44. Culy, C.R., N. Bhana, and G.L. Plosker, *Ondansetron: a review of its use as an antiemetic in children.* Paediatr Drugs, 2001. **3**(6): p. 441-79.

45. Graham-Pole, J., et al., *Antiemetics in children receiving cancer chemotherapy: a double-blind prospective randomized study comparing metoclopramide with chlorpromazine.* J Clin Oncol, 1986. **4**(7): p. 1110-3.
46. Launois, S., et al., *Hiccup in adults: an overview.* Eur Respir J, 1993. **6**(4): p. 563-75.
47. Lewis, J.H., *Hiccups: causes and cures.* J Clin Gastroenterol, 1985. **7**(6): p. 539-52.
48. Lipsky, M.S., *Chronic hiccups.* Am Fam Physician, 1986. **34**(5): p. 173-7.
49. Roila, F., M. Aapro, and A. Stewart, *Optimal selection of antiemetics in children receiving cancer chemotherapy.* Support Care Cancer, 1998. **6**(3): p. 215-20.
50. Williamson, B.W. and I.M. MacIntyre, *Management of intractable hiccup.* Br Med J, 1977. **2**(6085): p. 501-3.
51. MartindaleOnline, *The Complete Drug Reference*, S.C. Sweetman, Editor., Pharmaceutical Press.
52. Ashton, H., *Guidelines for the rational use of benzodiazepines. When and what to use.* Drugs, 1994. **48**(1): p. 25-40.
53. Larsson, P., et al., *Oral bioavailability of clonidine in children.* Paediatr Anaesth, 2011. **21**(3): p. 335-40.
54. Lambert, P., et al., *Clonidine premedication for postoperative analgesia in children.* Cochrane Database Syst Rev, 2014. **1**: p. CD009633.
55. Dahmani, S., et al., *Premedication with clonidine is superior to benzodiazepines. A meta analysis of published studies.* Acta Anaesthesiol Scand, 2010. **54**(4): p. 397-402.
56. Bergendahl, H., P.A. Lonnqvist, and S. Eksborg, *Clonidine in paediatric anaesthesia: review of the literature and comparison with benzodiazepines for premedication.* Acta Anaesthesiol Scand, 2006. **50**(2): p. 135-43.
57. Mitra, S., S. Kazal, and L.K. Anand, *Intranasal clonidine vs. midazolam as premedication in children: a randomized controlled trial.* Indian Pediatr, 2014. **51**(2): p. 113-8.
58. Mukherjee, A., *Characterization of alpha 2-adrenergic receptors in human platelets by binding of a radioactive ligand [3H]yohimbine.* Biochim Biophys Acta, 1981. **676**(2): p. 148-54.
59. Freeman, K.O., et al., *Analgesia for paediatric tonsillectomy and adenoidectomy with intramuscular clonidine.* Paediatr Anaesth, 2002. **12**(7): p. 617-20.

60. Arenas-Lopez, S., et al., *Use of oral clonidine for sedation in ventilated paediatric intensive care patients.* Intensive Care Med, 2004. **30**(8): p. 1625-9.
61. Ambrose, C., et al., *Intravenous clonidine infusion in critically ill children: dose-dependent sedative effects and cardiovascular stability.* Br J Anaesth, 2000. **84**(6): p. 794-6.
62. Honey, B.L., et al., *Alpha2-receptor agonists for treatment and prevention of iatrogenic opioid abstinence syndrome in critically ill patients.* Ann Pharmacother, 2009. **43**(9): p. 1506-11.
63. Schnabel, A., et al., *Efficacy and safety of clonidine as additive for caudal regional anesthesia: a quantitative systematic review of randomized controlled trials.* Paediatr Anaesth, 2011. **21**(12): p. 1219-30.
64. Allen, N.M., et al., *Status dystonicus: a practice guide.* Dev Med Child Neurol, 2014. **56**(2): p. 105-12.
65. Lubsch, L., et al., *Oral baclofen and clonidine for treatment of spasticity in children.* J Child Neurol, 2006. **21**(12): p. 1090-2.
66. Nguyen, M., et al., *A review of the use of clonidine as a sleep aid in the child and adolescent population.* Clin Pediatr (Phila), 2014. **53**(3): p. 211-6.
67. Potts, A.L., et al., *Clonidine disposition in children; a population analysis.* Paediatr Anaesth, 2007. **17**(10): p. 924-33.
68. Smith, H.S., *Opioid metabolism.* Mayo Clin Proc, 2009. **84**(7): p. 613-24.
69. Williams, D.G., A. Patel, and R.F. Howard, *Pharmacogenetics of codeine metabolism in an urban population of children and its implications for analgesic reliability.* Br J Anaesth, 2002. **89**(6): p. 839-45.
70. Drake, R., et al., *Impact of an antiemetic protocol on postoperative nausea and vomiting in children.* Paediatr Anaesth, 2001. **11**(1): p. 85-91.
71. Krach, L.E., *Pharmacotherapy of spasticity: oral medications and intrathecal baclofen.* J Child Neurol, 2001. **16**(1): p. 31-6.
72. Pinder, R.M., et al., *Dantrolene sodium: a review of its pharmacological properties and therapeutic efficacy in spasticity.* Drugs, 1977. **13**(1): p. 3-23.
73. Dupuis, L.L., R. Lau, and M.L. Greenberg, *Delayed nausea and vomiting in children receiving antineoplastics.* Med Pediatr Oncol, 2001. **37**(2): p. 115-21.
74. de Vries, M.A., et al., *Effect of dexamethasone on quality of life in children with acute lymphoblastic leukaemia: a prospective observational study.* Health Qual Life Outcomes, 2008. **6**(1): p. 103.

75. Tramer, M.R., *[Prevention and treatment of postoperative nausea and vomiting in children. An evidence-based approach]*. Ann Fr Anesth Reanim, 2007. **26**(6): p. 529-34.
76. Dupuis, L.L., et al., *Guideline for the prevention of acute nausea and vomiting due to antineoplastic medication in pediatric cancer patients.* Pediatric Blood and Cancer, 2013. **60**(7): p. 1073-1082.
77. Hewitt, M., et al., *Opioid use in palliative care of children and young people with cancer.* J Pediatr, 2008. **152**(1): p. 39-44.
78. Grimshaw, D., et al., *Subcutaneous midazolam, diamorphine and hyoscine infusion in palliative care of a child with neurodegenerative disease.* Child Care Health Dev, 1995. **21**(6): p. 377-81.
79. Camfield, P.R., *Buccal midazolam and rectal diazepam for treatment of prolonged seizures in childhood and adolescence: a randomised trial.* J Pediatr, 1999. **135**(3): p. 398-9.
80. Mathew, A., et al., *The efficacy of diazepam in enhancing motor function in children with spastic cerebral palsy.* J Trop Pediatr, 2005. **51**(2): p. 109-13.
81. Mitchell, W.G., *Status epilepticus and acute repetitive seizures in children, adolescents, and young adults: etiology, outcome, and treatment.* Epilepsia, 1996. **37 Suppl 1**: p. S74-80.
82. O'Dell, C. and K. O'Hara, *School nurses' experience with administration of rectal diazepam gel for seizures.* J Sch Nurs, 2007. **23**(3): p. 166-9.
83. O'Dell, C., et al., *Emergency management of seizures in the school setting.* J Sch Nurs, 2007. **23**(3): p. 158-65.
84. Srivastava, M. and D. Walsh, *Diazepam as an adjuvant analgesic to morphine for pain due to skeletal muscle spasm.* Support Care Cancer, 2003. **11**(1): p. 66-9.
85. Cinquetti, M., P. Bonetti, and P. Bertamini, *[Current role of antidopaminergic drugs in pediatrics].* Pediatr Med Chir, 2000. **22**(1): p. 1-7.
86. *Domperidone: an alternative to metoclopramide.* Drug Ther Bull, 1988. **26**(15): p. 59-60.
87. Demol, P., H.J. Ruoff, and T.R. Weihrauch, *Rational pharmacotherapy of gastrointestinal motility disorders.* Eur J Pediatr, 1989. **148**(6): p. 489-95.
88. Keady, S., *Update on drugs for gastro-oesophageal reflux disease.* Arch Dis Child Educ Pract Ed, 2007. **92**(4): p. ep114-8.

89. Pritchard, D.S., N. Baber, and T. Stephenson, *Should domperidone be used for the treatment of gastro-oesophageal reflux in children? Systematic review of randomized controlled trials in children aged 1 month to 11 years old.* Br J Clin Pharmacol, 2005. **59**(6): p. 725-9.

90. MHRA, *Domperidone: small risk of serious ventricular arrhythmia and sudden cardiac death.* 2012. p. A2.

91. Gubbay, A. and K. Langdon, *'Effectiveness of sedation using nitrous oxide compared with enteral midazolam for botulinum toxin A injections in children'.* Dev Med Child Neurol, 2009. **51**(6): p. 491-2; author reply 492.

92. Bellomo-Brandao, M.A., E.F. Collares, and E.A. da-Costa-Pinto, *Use of erythromycin for the treatment of severe chronic constipation in children.* Braz J Med Biol Res, 2003. **36**(10): p. 1391-6.

93. Novak, P.H., et al., *Acute drug prescribing to children on chronic antiepilepsy therapy and the potential for adverse drug interactions in primary care.* Br J Clin Pharmacol, 2005. **59**(6): p. 712-7.

94. Tsoukas, C., et al., *Evaluation of the efficacy and safety of etoricoxib in the treatment of hemophilic arthropathy.* Blood, 2006. **107**(5): p. 1785-90.

95. Grape, S., et al., *Formulations of fentanyl for the management of pain.* Drugs. **70**(1): p. 57-72.

96. Cappelli, C., et al., *[Transdermal Fentanyl: news in oncology.].* Clin Ter, 2008. **159**(4): p. 257-260.

97. Weschules, D.J., et al., *Toward evidence-based prescribing at end of life: a comparative analysis of sustained-release morphine, oxycodone, and transdermal fentanyl, with pain, constipation, and caregiver interaction outcomes in hospice patients.* Pain Med, 2006. **7**(4): p. 320-9.

98. Borland, M., et al., *A randomized controlled trial comparing intranasal fentanyl to intravenous morphine for managing acute pain in children in the emergency department.* Ann Emerg Med, 2007. **49**(3): p. 335-40.

99. Borland, M.L., I. Jacobs, and G. Geelhoed, *Intranasal fentanyl reduces acute pain in children in the emergency department: a safety and efficacy study.* Emerg Med (Fremantle), 2002. **14**(3): p. 275-80.

100. Drake, R., J. Longworth, and J.J. Collins, *Opioid rotation in children with cancer.* J Palliat Med, 2004. **7**(3): p. 419-22.

101. Friedrichsdorf, S.J. and T.I. Kang, *The management of pain in children with life-limiting illnesses.* Pediatr Clin North Am, 2007. **54**(5): p. 645-72, x.

102. Hunt, A., et al., *Transdermal fentanyl for pain relief in a paediatric palliative care population.* Palliat Med, 2001. **15**(5): p. 405-12.

103. Kanowitz, A., et al., *Safety and effectiveness of fentanyl administration for prehospital pain management.* Prehosp Emerg Care, 2006. **10**(1): p. 1-7.

104. Mercadante, S., et al., *Transmucosal fentanyl vs intravenous morphine in doses proportional to basal opioid regimen for episodic-breakthrough pain.* Br J Cancer, 2007. **96**(12): p. 1828-33.

105. Noyes, M. and H. Irving, *The use of transdermal fentanyl in pediatric oncology palliative care.* Am J Hosp Palliat Care, 2001. **18**(6): p. 411-6.

106. Weschules, D.J., et al., *Are newer, more expensive pharmacotherapy options associated with superior symptom control compared to less costly agents used in a collaborative practice setting?* Am J Hosp Palliat Care, 2006. **23**(2): p. 135-49.

107. Harlos, M.S., et al., *Intranasal fentanyl in the palliative care of newborns and infants.* J Pain Symptom Manage, 2013. **46**(2): p. 265-74.

108. Pienaar, E.D., T. Young, and H. Holmes, *Interventions for the prevention and management of oropharyngeal candidiasis associated with HIV infection in adults and children.* Cochrane Database Syst Rev, 2006. **3**: p. CD003940.

109. Pfizer. *DIFLUCAN U.S. Physician Prescribing Information* 2014; Available from: http://www.pfizer.com/products/product-detail/diflucan.

110. Emslie, G.J., et al., *Fluoxetine Versus Placebo in Preventing Relapse of Major Depression in Children and Adolescents.* Am J Psychiatry, 2008.

111. Birmaher, B., et al., *Fluoxetine for the treatment of childhood anxiety disorders.* J Am Acad Child Adolesc Psychiatry, 2003. **42**(4): p. 415-23.

112. Hetrick, S., et al., *Selective serotonin reuptake inhibitors (SSRIs) for depressive disorders in children and adolescents.* Cochrane Database Syst Rev, 2007(3): p. CD004851.

113. Jick, H., J.A. Kaye, and S.S. Jick, *Antidepressants and the risk of suicidal behaviors.* Jama, 2004. **292**(3): p. 338-43.

114. Millet, B., et al., *Obsessive-compulsive disorder: evaluation of clinical and biological circadian parameters during fluoxetine treatment.* Psychopharmacology (Berl), 1999. **146**(3): p. 268-74.

115. Monteleone, P., et al., *Plasma melatonin and cortisol circadian patterns in patients with obsessive-compulsive disorder before and after fluoxetine treatment.* Psychoneuroendocrinology, 1995. **20**(7): p. 763-70.

116. Roth, D., et al., *Depressing research.* Lancet, 2004. **363**(9426): p. 2087.
117. Whittington, C.J., et al., *Selective serotonin reuptake inhibitors in childhood depression: systematic review of published versus unpublished data.* Lancet, 2004. **363**(9418): p. 1341-5.
118. Caraceni, A., et al., *Gabapentin for neuropathic cancer pain: a randomized controlled trial from the Gabapentin Cancer Pain Study Group.* J Clin Oncol, 2004. **22**(14): p. 2909-17.
119. Butkovic, D., S. Toljan, and B. Mihovilovic-Novak, *Experience with gabapentin for neuropathic pain in adolescents: report of five cases.* Paediatr Anaesth, 2006. **16**(3): p. 325-9.
120. Pfizer. *NEURONTIN U.S. Physician Prescribing Information.* 2014; Available from: http://www.pfizer.com/products/product-detail/neurontin.
121. Wee, B. and R. Hillier, *Interventions for noisy breathing in patients near to death.* Cochrane Database Syst Rev, 2008(1): p. CD005177.
122. Back, I.N., et al., *A study comparing hyoscine hydrobromide and glycopyrrolate in the treatment of death rattle.* Palliat Med, 2001. **15**(4): p. 329-36.
123. Bennett, M., et al., *Using anti-muscarinic drugs in the management of death rattle: evidence-based guidelines for palliative care.* Palliat Med, 2002. **16**(5): p. 369-74.
124. Dumortier, G., et al., *[Prescription of psychotropic drugs in paediatry: approved indications and therapeutic perspectives].* Encephale, 2005. **31**(4 Pt 1): p. 477-89.
125. Breitbart, W., et al., *A double-blind trial of haloperidol, chlorpromazine, and lorazepam in the treatment of delirium in hospitalized AIDS patients.* Am J Psychiatry, 1996. **153**(2): p. 231-7.
126. Breitbart, W. and D. Strout, *Delirium in the terminally ill.* Clin Geriatr Med, 2000. **16**(2): p. 357-72.
127. Negro, S., et al., *Physical compatibility and in vivo evaluation of drug mixtures for subcutaneous infusion to cancer patients in palliative care.* Support Care Cancer, 2002. **10**(1): p. 65-70.
128. Saito, T. and S. Shinno, *[How we have treated and cared patients with Duchenne muscular dystrophy and severe congestive heart failure].* No To Hattatsu, 2005. **37**(4): p. 281-6.
129. Bell, R.F., et al., *Controlled clinical trials in cancer pain. How controlled should they be? A qualitative systematic review.* Br J Cancer, 2006.

130. Quigley, C. and P. Wiffen, *A systematic review of hydromorphone in acute and chronic pain.* J Pain Symptom Manage, 2003. **25**(2): p. 169-78.
131. Titchen, T., N. Cranswick, and S. Beggs, *Adverse drug reactions to nonsteroidal anti-inflammatory drugs, COX-2 inhibitors and paracetamol in a paediatric hospital.* Br J Clin Pharmacol, 2005. **59**(6): p. 718-23.
132. Anderson, B.J. and G.M. Palmer, *Recent developments in the pharmacological management of pain in children.* Curr Opin Anaesthesiol, 2006. **19**(3): p. 285-92.
133. Anghelescu, D.L. and L.L. Oakes, *Ketamine use for reduction of opioid tolerance in a 5-year-old girl with end-stage abdominal neuroblastoma.* J Pain Symptom Manage, 2005. **30**(1): p. 1-3.
134. Campbell-Fleming, J.M. and A. Williams, *The use of ketamine as adjuvant therapy to control severe pain.* Clin J Oncol Nurs, 2008. **12**(1): p. 102-7.
135. Legge, J., N. Ball, and D.P. Elliott, *The potential role of ketamine in hospice analgesia: a literature review.* Consult Pharm, 2006. **21**(1): p. 51-7.
136. Tsui, B.C., et al., *Intravenous ketamine infusion as an adjuvant to morphine in a 2-year-old with severe cancer pain from metastatic neuroblastoma.* J Pediatr Hematol Oncol, 2004. **26**(10): p. 678-80.
137. Fitzgibbon, E.J., et al., *Low dose ketamine as an analgesic adjuvant in difficult pain syndromes: a strategy for conversion from parenteral to oral ketamine.* J Pain Symptom Manage, 2002. **23**(2): p. 165-70.
138. Bell, R., C. Eccleston, and E. Kalso, *Ketamine as an adjuvant to opioids for cancer pain.* Cochrane Database Syst Rev, 2003(1): p. CD003351.
139. Klepstad, P., et al., *Long-term treatment with ketamine in a 12-year-old girl with severe neuropathic pain caused by a cervical spinal tumor.* J Pediatr Hematol Oncol, 2001. **23**(9): p. 616-9.
140. Benitez-Rosario, M.A., et al., *A strategy for conversion from subcutaneous to oral ketamine in cancer pain patients: effect of a 1:1 ratio.* J Pain Symptom Manage, 2011. **41**(6): p. 1098-105.
141. Chen, C.H., et al., *Ketamine-snorting associated cystitis.* J Formos Med Assoc, 2011. **110**(12): p. 787-91.
142. Shahani, R., et al., *Ketamine-associated ulcerative cystitis: a new clinical entity.* Urology, 2007. **69**(5): p. 810-2.
143. Aldrink, J.H., et al., *Safety of ketorolac in surgical neonates and infants 0 to 3 months old.* J Pediatr Surg, 2011. **46**(6): p. 1081-5.

144. Cohen, M.N., et al., *Pharmacokinetics of single-dose intravenous ketorolac in infants aged 2-11 months.* Anesth Analg, 2011. **112**(3): p. 655-60.
145. Zuppa, A.F., et al., *Population pharmacokinetics of ketorolac in neonates and young infants.* Am J Ther, 2009. **16**(2): p. 143-6.
146. Hong, J.Y., et al., *Fentanyl sparing effects of combined ketorolac and acetaminophen for outpatient inguinal hernia repair in children.* J Urol, 2010. **183**(4): p. 1551-5.
147. Jo, Y.Y., et al., *Ketorolac or fentanyl continuous infusion for post-operative analgesia in children undergoing ureteroneocystostomy.* Acta Anaesthesiol Scand, 2011. **55**(1): p. 54-9.
148. Keidan, I., et al., *Intraoperative ketorolac is an effective substitute for fentanyl in children undergoing outpatient adenotonsillectomy.* Paediatr Anaesth, 2004. **14**(4): p. 318-23.
149. Moreno, M., F.J. Castejon, and M.A. Palacio, *Patient-controlled analgesia with ketorolac in pediatric surgery.* J Physiol Biochem, 2000. **56**(3): p. 209-16.
150. Shende, D. and K. Das, *Comparative effects of intravenous ketorolac and pethidine on perioperative analgesia and postoperative nausea and vomiting (PONV) for paediatric strabismus surgery.* Acta Anaesthesiol Scand, 1999. **43**(3): p. 265-9.
151. Chiaretti, A., et al., *[Analgesic efficacy of ketorolac and fentanyl in pediatric intensive care].* Pediatr Med Chir, 1997. **19**(6): p. 419-24.
152. Forrest, J.B., E.L. Heitlinger, and S. Revell, *Ketorolac for postoperative pain management in children.* Drug Saf, 1997. **16**(5): p. 309-29.
153. Gillis, J.C. and R.N. Brogden, *Ketorolac. A reappraisal of its pharmacodynamic and pharmacokinetic properties and therapeutic use in pain management.* Drugs, 1997. **53**(1): p. 139-88.
154. Urganci, N., B. Akyildiz, and T.B. Polat, *A comparative study: the efficacy of liquid paraffin and lactulose in management of chronic functional constipation.* Pediatr Int, 2005. **47**(1): p. 15-9.
155. Candy, D.C., D. Edwards, and M. Geraint, *Treatment of faecal impaction with polyethelene glycol plus electrolytes (PGE + E) followed by a double-blind comparison of PEG + E versus lactulose as maintenance therapy.* J Pediatr Gastroenterol Nutr, 2006. **43**(1): p. 65-70.
156. Lee-Robichaud, H., et al., *Lactulose versus Polyethylene Glycol for Chronic Constipation.* Cochrane Database Syst Rev, 2010(7): p. CD007570.

157. Orenstein, S.R., et al., *Multicenter, double-blind, randomized, placebo-controlled trial assessing the efficacy and safety of proton pump inhibitor lansoprazole in infants with symptoms of gastroesophageal reflux disease.* J Pediatr, 2009. **154**(4): p. 514-520 e4.
158. Khoshoo, V. and P. Dhume, *Clinical response to 2 dosing regimens of lansoprazole in infants with gastroesophageal reflux.* J Pediatr Gastroenterol Nutr, 2008. **46**(3): p. 352-4.
159. Gremse, D., et al., *Pharmacokinetics and pharmacodynamics of lansoprazole in children with gastroesophageal reflux disease.* J Pediatr Gastroenterol Nutr, 2002. **35 Suppl 4**: p. S319-26.
160. Tolia, V., et al., *Efficacy of lansoprazole in the treatment of gastroesophageal reflux disease in children.* J Pediatr Gastroenterol Nutr, 2002. **35 Suppl 4**: p. S308-18.
161. Tolia, V., et al., *Safety of lansoprazole in the treatment of gastroesophageal reflux disease in children.* J Pediatr Gastroenterol Nutr, 2002. **35 Suppl 4**: p. S300-7.
162. Tolia, V. and Y. Vandenplas, *Systematic review: the extra-oesophageal symptoms of gastro-oesophageal reflux disease in children.* Aliment Pharmacol Ther, 2009. **29**(3): p. 258-72.
163. Heyman, M.B., et al., *Pharmacokinetics and pharmacodynamics of lansoprazole in children 13 to 24 months old with gastroesophageal reflux disease.* J Pediatr Gastroenterol Nutr, 2007. **44**(1): p. 35-40.
164. Tran, A., et al., *Pharmacokinetic-pharmacodynamic study of oral lansoprazole in children.* Clin Pharmacol Ther, 2002. **71**(5): p. 359-67.
165. Gunasekaran, T., et al., *Lansoprazole in adolescents with gastroesophageal reflux disease: pharmacokinetics, pharmacodynamics, symptom relief efficacy, and tolerability.* J Pediatr Gastroenterol Nutr, 2002. **35 Suppl 4**: p. S327-35.
166. Zhang, W., et al., *Age-dependent pharmacokinetics of lansoprazole in neonates and infants.* Paediatr Drugs, 2008. **10**(4): p. 265-74.
167. Springer, M., et al., *Safety and pharmacodynamics of lansoprazole in patients with gastroesophageal reflux disease aged <1 year.* Paediatr Drugs, 2008. **10**(4): p. 255-63.
168. Franco, M.T., et al., *Lansoprazole in the treatment of gastro-oesophageal reflux disease in childhood.* Dig Liver Dis, 2000. **32**(8): p. 660-6.

169. Faure, C., et al., *Lansoprazole in children: pharmacokinetics and efficacy in reflux oesophagitis.* Aliment Pharmacol Ther, 2001. **15**(9): p. 1397-402.
170. Litalien, C., Y. Theoret, and C. Faure, *Pharmacokinetics of proton pump inhibitors in children.* Clin Pharmacokinet, 2005. **44**(5): p. 441-66.
171. Skinner, J. and A. Skinner, *Levomepromazine for nausea and vomiting in advanced cancer.* Hosp Med, 1999. **60**(8): p. 568-70.
172. O'Neill, J. and A. Fountain, *Levomepromazine (methotrimeprazine) and the last 48 hours.* Hosp Med, 1999. **60**(8): p. 564-7.
173. Hohl, C.M., et al., *Methotrimeprazine for the management of end-of-life symptoms in infants and children.* J Palliat Care, 2013. **29**(3): p. 178-85.
174. Hans, G., et al., *Management of neuropathic pain after surgical and non-surgical trauma with lidocaine 5% patches: study of 40 consecutive cases.* Curr Med Res Opin, 2009. **25**(11): p. 2737-43.
175. Garnock-Jones, K.P. and G.M. Keating, *Lidocaine 5% medicated plaster: a review of its use in postherpetic neuralgia.* Drugs, 2009. **69**(15): p. 2149-65.
176. *Lidocaine plasters for postherpetic neuralgia?* Drug Ther Bull, 2008. **46**(2): p. 14-6.
177. Binder, A., et al., *Topical 5% lidocaine (lignocaine) medicated plaster treatment for post-herpetic neuralgia: results of a double-blind, placebo-controlled, multinational efficacy and safety trial.* Clin Drug Investig, 2009. **29**(6): p. 393-408.
178. Hans, G., et al., *Efficacy and tolerability of a 5% lidocaine medicated plaster for the topical treatment of post-herpetic neuralgia: results of a long-term study.* Curr Med Res Opin, 2009. **25**(5): p. 1295-305.
179. Nalamachu, S., et al., *Influence of anatomic location of lidocaine patch 5% on effectiveness and tolerability for postherpetic neuralgia.* Patient Prefer Adherence, 2013. **7**: p. 551-7.
180. Karan, S., *Lomotil in diarrhoeal illnesses.* Arch Dis Child, 1979. **54**(12): p. 984.
181. Bala, K., S.S. Khandpur, and V.V. Gujral, *Evaluation of efficacy and safety of lomotil in acute diarrhoeas in children.* Indian Pediatr, 1979. **16**(10): p. 903-7.
182. Waterston, A.J., *Lomotil in diarrhoeal illnesses.* Arch Dis Child, 1980. **55**(7): p. 577-8.

183. Li, S.T., D.C. Grossman, and P. Cummings, *Loperamide therapy for acute diarrhea in children: systematic review and meta-analysis.* PLoS Med, 2007. **4**(3): p. e98.
184. Kaplan, M.A., et al., *A multicenter randomized controlled trial of a liquid loperamide product versus placebo in the treatment of acute diarrhea in children.* Clin Pediatr (Phila), 1999. **38**(10): p. 579-91.
185. Omar, M.I. and C.E. Alexander, *Drug treatment for faecal incontinence in adults.* Cochrane Database Syst Rev, 2013. **6**: p. CD002116.
186. Burtles, R. and B. Astley, *Lorazepam in children. A double-blind trial comparing lorazepam, diazepam, trimeprazine and placebo.* Br J Anaesth, 1983. **55**(4): p. 275-9.
187. Hanson, S. and N. Bansal, *The clinical effectiveness of Movicol in children with severe constipation: an outcome audit.* Paediatr Nurs, 2006. **18**(2): p. 24-8.
188. NICE. *Constipation in Children and Young People.* 2010 May 2010]; CG99 [Available from: http://guidance.nice.org.uk/CG99.
189. Braam, W., et al., *Melatonin treatment in individuals with intellectual disability and chronic insomnia: a randomized placebo-controlled study.* J Intellect Disabil Res, 2008. **52**(Pt 3): p. 256-64.
190. Andersen, I.M., et al., *Melatonin for insomnia in children with autism spectrum disorders.* J Child Neurol, 2008. **23**(5): p. 482-5.
191. Guerrero, J.M., et al., *Impairment of the melatonin rhythm in children with Sanfilippo syndrome.* J Pineal Res, 2006. **40**(2): p. 192-3.
192. Gupta, R. and J. Hutchins, *Melatonin: a panacea for desperate parents? (Hype or truth).* Arch Dis Child, 2005. **90**(9): p. 986-7.
193. Ivanenko, A., et al., *Melatonin in children and adolescents with insomnia: a retrospective study.* Clin Pediatr (Phila), 2003. **42**(1): p. 51-8.
194. Mariotti, P., et al., *Sleep disorders in Sanfilippo syndrome: a polygraphic study.* Clin Electroencephalogr, 2003. **34**(1): p. 18-22.
195. Masters, K.J., *Melatonin for sleep problems.* J Am Acad Child Adolesc Psychiatry, 1996. **35**(6): p. 704.
196. Owens, J.A., C.L. Rosen, and J.A. Mindell, *Medication use in the treatment of pediatric insomnia: results of a survey of community-based pediatricians.* Pediatrics, 2003. **111**(5 Pt 1): p. e628-35.

197. Paavonen, E.J., et al., *Effectiveness of melatonin in the treatment of sleep disturbances in children with Asperger disorder.* J Child Adolesc Psychopharmacol, 2003. **13**(1): p. 83-95.

198. Smits, M.G., et al., *Melatonin for chronic sleep onset insomnia in children: a randomized placebo-controlled trial.* J Child Neurol, 2001. **16**(2): p. 86-92.

199. Smits, M.G., et al., *Melatonin improves health status and sleep in children with idiopathic chronic sleep-onset insomnia: a randomized placebo-controlled trial.* J Am Acad Child Adolesc Psychiatry, 2003. **42**(11): p. 1286-93.

200. van der Heijden, K.B., et al., *Prediction of melatonin efficacy by pretreatment dim light melatonin onset in children with idiopathic chronic sleep onset insomnia.* J Sleep Res, 2005. **14**(2): p. 187-94.

201. Van der Heijden, K.B., et al., *Effect of melatonin on sleep, behavior, and cognition in ADHD and chronic sleep-onset insomnia.* J Am Acad Child Adolesc Psychiatry, 2007. **46**(2): p. 233-41.

202. Wasdell, M.B., et al., *A randomized, placebo-controlled trial of controlled release melatonin treatment of delayed sleep phase syndrome and impaired sleep maintenance in children with neurodevelopmental disabilities.* J Pineal Res, 2008. **44**(1): p. 57-64.

203. Zhdanova, I.V., *Melatonin as a hypnotic: pro.* Sleep Med Rev, 2005. **9**(1): p. 51-65.

204. Zucconi, M. and O. Bruni, *Sleep disorders in children with neurologic diseases.* Semin Pediatr Neurol, 2001. **8**(4): p. 258-75.

205. Gringras, P., et al., *Melatonin for sleep problems in children with neurodevelopmental disorders: randomised double masked placebo controlled trial.* BMJ, 2012. **345**: p. e6664.

206. Ferracioli-Oda, E., A. Qawasmi, and M.H. Bloch, *Meta-analysis: melatonin for the treatment of primary sleep disorders.* PLoS One, 2013. **8**(5): p. e63773.

207. Dickman, A. and J. Schneider, *The Syringe Driver. Continuous Infusions in Palliative Care.* 3rd ed. 2011: Oxford University Press.

208. Benitez-Rosario, M.A., et al., *Morphine-methadone opioid rotation in cancer patients: analysis of dose ratio predicting factors.* J Pain Symptom Manage, 2009. **37**(6): p. 1061-8.

209. Bruera, E., et al., *Methadone versus morphine as a first-line strong opioid for cancer pain: a randomized, double-blind study.* J Clin Oncol, 2004. **22**(1): p. 185-92.

210. Berens, R.J., et al., *A prospective evaluation of opioid weaning in opioid-dependent pediatric critical care patients.* Anesth Analg, 2006. **102**(4): p. 1045-50.

211. Colvin, L., K. Forbes, and M. Fallon, *Difficult pain.* Bmj, 2006. **332**(7549): p. 1081-3.

212. Dale, O., P. Sheffels, and E.D. Kharasch, *Bioavailabilities of rectal and oral methadone in healthy subjects.* Br J Clin Pharmacol, 2004. **58**(2): p. 156-62.

213. Davies, D., D. DeVlaming, and C. Haines, *Methadone analgesia for children with advanced cancer.* Pediatr Blood Cancer, 2008. **51**(3): p. 393-7.

214. Ripamonti, C. and M. Bianchi, *The use of methadone for cancer pain.* Hematol Oncol Clin North Am, 2002. **16**(3): p. 543-55.

215. Weschules, D.J. and K.T. Bain, *A systematic review of opioid conversion ratios used with methadone for the treatment of pain.* Pain Med, 2008. **9**(5): p. 595-612.

216. Weschules, D.J., et al., *Methadone and the hospice patient: prescribing trends in the home-care setting.* Pain Med, 2003. **4**(3): p. 269-76.

217. Heppe, D.B., M.C. Haigney, and M.J. Krantz, *The effect of oral methadone on the QTc interval in advanced cancer patients: a prospective pilot study.* J Palliat Med. **13**(6): p. 638-9.

218. Mercadante, S., P. Ferrera, and E. Arcuri, *The use of fentanyl buccal tablets as breakthrough medication in patients receiving chronic methadone therapy: an open label preliminary study.* Support Care Cancer.

219. Mercadante, S., et al., *Changes of QTc interval after opioid switching to oral methadone.* Support Care Cancer, 2013. **21**(12): p. 3421-4.

220. www.palliativedrugs.com, *Methylnaltrexone.* 2010.

221. Rodriques A et al, *Methylnaltrexone for Opiod-Induced Constipation in Pediatric Oncology Patients. Pediatr Blood Cancer.* Pediatr Blood Cancer, 2013. **Jun1**(4).

222. Laubisch, J.E. and J.N. Baker, *Methylnaltrexone use in a seventeen-month-old female with progressive cancer and rectal prolapse.* J Palliat Med, 2013. **16**(11): p. 1486-8.

223. Garten, L. and C. Buhrer, *Reversal of morphine-induced urinary retention after methylnaltrexone.* Arch Dis Child Fetal Neonatal Ed, 2012. **97**(2): p. F151-3.

224. Garten, L., P. Degenhardt, and C. Buhrer, *Resolution of opioid-induced postoperative ileus in a newborn infant after methylnaltrexone.* J Pediatr Surg, 2011. **46**(3): p. e13-5.

225. Kissling, K.T., L.R. Mohassel, and J. Heintz, *Methylnaltrexone for opioid-induced constipation in a pediatric oncology patient.* J Pain Symptom Manage, 2012. **44**(1): p. e1-3.

226. Lee, J.M. and J. Mooney, *Methylnaltrexone in treatment of opioid-induced constipation in a pediatric patient.* Clin J Pain, 2012. **28**(4): p. 338-41.

227. Madanagopolan, N., *Metoclopramide in hiccup.* Curr Med Res Opin, 1975. **3**(6): p. 371-4.

228. Alhashimi, D., H. Alhashimi, and Z. Fedorowicz, *Antiemetics for reducing vomiting related to acute gastroenteritis in children and adolescents.* Cochrane Database Syst Rev, 2006. **3**: p. CD005506.

229. Craig, W.R., et al., *Metoclopramide, thickened feedings, and positioning for gastro-oesophageal reflux in children under two years.* The Cochrane Database of Systematic Reviews, 2004. **2004**(3.).

230. Yis, U., et al., *Metoclopramide induced dystonia in children: two case reports.* Eur J Emerg Med, 2005. **12**(3): p. 117-9.

231. EMA, *European Medicines Agency recommends changes to the use of metoclopramide.* 2013.

232. Trindade, L.C., et al., *Evaluation of topical metronidazole in the healing wounds process: an experimental study.* Rev Col Bras Cir, 2010. **37**(5): p. 358-63.

233. Collins, C.D., S. Cookinham, and J. Smith, *Management of oropharyngeal candidiasis with localized oral miconazole therapy: efficacy, safety, and patient acceptability.* Patient Prefer Adherence, 2011. **5**: p. 369-74.

234. Mpimbaza, A., et al., *Comparison of buccal midazolam with rectal diazepam in the treatment of prolonged seizures in Ugandan children: a randomized clinical trial.* Pediatrics, 2008. **121**(1): p. e58-64.

235. Scott, R.C., F.M. Besag, and B.G. Neville, *Buccal midazolam and rectal diazepam for treatment of prolonged seizures in childhood and adolescence: a randomised trial.* Lancet, 1999. **353**(9153): p. 623-6.

236. Castro Conde, J.R., et al., *Midazolam in neonatal seizures with no response to phenobarbital.* Neurology, 2005. **64**(5): p. 876-9.

237. Harte, G.J., et al., *Haemodynamic responses and population pharmacokinetics of midazolam following administration to ventilated, preterm neonates.* J Paediatr Child Health, 1997. **33**(4): p. 335-8.
238. Lee, T.C., et al., *Population pharmacokinetic modeling in very premature infants receiving midazolam during mechanical ventilation: midazolam neonatal pharmacokinetics.* Anesthesiology, 1999. **90**(2): p. 451-7.
239. Hu, K.C., et al., *Continuous midazolam infusion in the treatment of uncontrollable neonatal seizures.* Acta Paediatr Taiwan, 2003. **44**(5): p. 279-81.
240. Berde, C.B. and N.F. Sethna, *Drug therapy - Analgesics for the treatment of pain in children.* New England Journal of Medicine, 2002. **347**(14): p. 1094-1103.
241. Boyle, E.M., et al., *Assessment of persistent pain or distress and adequacy of analgesia in preterm ventilated infants.* Pain, 2006. **124**(1-2): p. 87-91.
242. Cohen, S.P. and T.C. Dawson, *Nebulized morphine as a treatment for dyspnea in a child with cystic fibrosis.* Pediatrics, 2002. **110**(3): p. e38.
243. Dougherty, M. and M.R. DeBaun, *Rapid increase of morphine and benzodiazepine usage in the last three days of life in children with cancer is related to neuropathic pain.* J Pediatr, 2003. **142**(4): p. 373-6.
244. Flogegard, H. and G. Ljungman, *Characteristics and adequacy of intravenous morphine infusions in children in a paediatric oncology setting.* Med Pediatr Oncol, 2003. **40**(4): p. 233-8.
245. Hain, R.D., et al., *Strong opioids in pediatric palliative medicine.* Paediatr Drugs, 2005. **7**(1): p. 1-9.
246. Hall, R.W., et al., *Morphine, Hypotension, and Adverse Outcomes Among Preterm Neonates: Who's to Blame? Secondary Results From the NEOPAIN Trial.* Pediatrics, 2005. **115**(5): p. 1351-1359.
247. Lundeberg, S., et al., *Perception of pain following rectal administration of morphine in children: a comparison of a gel and a solution.* Paediatr Anaesth, 2006. **16**(2): p. 164-9.
248. Miser, A.W., et al., *Continuous subcutaneous infusion of morphine in children with cancer.* Am J Dis Child, 1983. **137**(4): p. 383-5.
249. Nahata, M.C., et al., *Analgesic plasma concentrations of morphine in children with terminal malignancy receiving a continuous subcutaneous infusion of morphine sulfate to control severe pain.* Pain, 1984. **18**(2): p. 109-14.

250. Sittl, R. and R. Richter, *[Cancer pain therapy in children and adolescents using morphine]*. Anaesthesist, 1991. **40**(2): p. 96-9.
251. Van Hulle Vincent, C. and M.J. Denyes, *Relieving children's pain: nurses' abilities and analgesic administration practices*. J Pediatr Nurs, 2004. **19**(1): p. 40-50.
252. Viola, R., et al., *The management of dyspnea in cancer patients: a systematic review*. Support Care Cancer, 2008.
253. Wiffen, P.J. and H.J. McQuay, *Oral morphine for cancer pain*. Cochrane Database Syst Rev, 2007(4): p. CD003868.
254. Zeppetella, G., J. Paul, and M.D. Ribeiro, *Analgesic efficacy of morphine applied topically to painful ulcers*. J Pain Symptom Manage, 2003. **25**(6): p. 555-8.
255. Zernikow, B. and G. Lindena, *Long-acting morphine for pain control in paediatric oncology*. Medical & Pediatric Oncology, 2001. **36**(4): p. 451-458.
256. Zernikow, B., et al., *Paediatric cancer pain management using the WHO analgesic ladder--results of a prospective analysis from 2265 treatment days during a quality improvement study*. Eur J Pain, 2006. **10**(7): p. 587-95.
257. Kaiko, R.F., et al., *The bioavailability of morphine in controlled-release 30-mg tablets per rectum compared with immediate-release 30-mg rectal suppositories and controlled-release 30-mg oral tablets*. Pharmacotherapy, 1992. **12**(2): p. 107-13.
258. Wilkinson, T.J., et al., *Pharmacokinetics and efficacy of rectal versus oral sustained-release morphine in cancer patients*. Cancer Chemother Pharmacol, 1992. **31**(3): p. 251-4.
259. Campbell, W.I., *Rectal controlled-release morphine: plasma levels of morphine and its metabolites following the rectal administration of MST Continus 100 mg*. J Clin Pharm Ther, 1996. **21**(2): p. 65-71.
260. Tofil, N.M., et al., *The use of enteral naloxone to treat opioid-induced constipation in a pediatric intensive care unit*. Pediatr Crit Care Med, 2006. **7**(3): p. 252-4.
261. Liu, M. and E. Wittbrodt, *Low-dose oral naloxone reverses opioid-induced constipation and analgesia*. J Pain Symptom Manage, 2002. **23**(1): p. 48-53.
262. Glenny, A.M., et al., *A survey of current practice with regard to oral care for children being treated for cancer*. Eur J Cancer, 2004. **40**(8): p. 1217-24.

263. Sassano-Higgins S et al, *Olanzapine reduces delirium symptoms in the critically ill pediatric patient.* J Pediatr Intensive Care, 2013. **2**(2): p. 49-54.
264. Beckwitt-Turkel S et al, *The diagnosis and management of delirium in infancy.* J Child Adolesc Psychopharmacol, 2013. **23**(5): p. 352-56.
265. Turkel SB et al, *Atypical antipsychotic medications to control symptoms of delirium in children and adolescents.* J Child Adolesc Psychopharmacol, 2012. **22**(2): p. 126-130.
266. Kaneishi, K., M. Kawabata, and T. Morita, *Olanzapine for the relief of nausea in patients with advanced cancer and incomplete bowel obstruction.* J Pain Symptom Manage, 2012. **44**(4): p. 604-7.
267. Kitada T et al, *Olanzapine as an antiemetic in intractable nausea and anorexia in patients with advanced hepatocellular carcinoma: 3 case series.* Acta Hepatolgica Japonica, 2009. **50**(3): p. 153-158.
268. Srivastava, M., et al., *Olanzapine as an antiemetic in refractory nausea and vomiting in advanced cancer.* J Pain Symptom Manage, 2003. **25**(6): p. 578-82.
269. Jackson, W.C. and L. Tavernier, *Olanzapine for intractable nausea in palliative care patients.* J Palliat Med, 2003. **6**(2): p. 251-5.
270. Passik, S.D., et al., *A pilot exploration of the antiemetic activity of olanzapine for the relief of nausea in patients with advanced cancer and pain.* J Pain Symptom Manage, 2002. **23**(6): p. 526-32.
271. Licup, N., *Olanzapine for nausea and vomiting.* Am J Hosp Palliat Care, 2010. **27**(6): p. 432-4.
272. Elsayem, A., et al., *Subcutaneous olanzapine for hyperactive or mixed delirium in patients with advanced cancer: a preliminary study.* J Pain Symptom Manage, 2010. **40**(5): p. 774-82.
273. Jackson KC et al, *Drug therapy for delirium in terminally ill adult patients.* Cochrane Database of Systematic Reviews, 2009.
274. Breitbart, W., A. Tremblay, and C. Gibson, *An open trial of olanzapine for the treatment of delirium in hospitalized cancer patients.* Psychosomatics, 2002. **43**(3): p. 175-82.
275. Khojainova, N., et al., *Olanzapine in the management of cancer pain.* J Pain Symptom Manage, 2002. **23**(4): p. 346-50.

276. Navari, R.M. and M.C. Brenner, *Treatment of cancer-related anorexia with olanzapine and megestrol acetate: a randomized trial.* Support Care Cancer, 2010. **18**(8): p. 951-6.

277. Gold, B.D., *Review article: epidemiology and management of gastro-oesophageal reflux in children.* Aliment Pharmacol Ther, 2004. **19 Suppl 1**: p. 22-7.

278. Chang, A.B., et al., *Gastro-oesophageal reflux treatment for prolonged non-specific cough in children and adults.* Cochrane Database Syst Rev, 2005(2): p. CD004823.

279. Simpson, T. and J. Ivey, *Pediatric management problems. GERD.* Pediatr Nurs, 2005. **31**(3): p. 214-5.

280. *5HT3-receptor antagonists as antiemetics in cancer.* Drug Ther Bull, 2005. **43**(8): p. 57-62.

281. Kyriakides, K., S.K. Hussain, and G.J. Hobbs, *Management of opioid-induced pruritus: a role for 5-HT3 antagonists?* Br J Anaesth, 1999. **82**(3): p. 439-41.

282. MHRA Drug Safety Update. *Ondansetron for intravenous use: dose-dependent QT interval prolongation – new posology.* 2013; July ; 6(12): :[Available from: http://www.mhra.gov.uk/Safetyinformation/DrugSafetyUpdate/CON296402.

283. Kokki, H., et al., *Comparison of oxycodone pharmacokinetics after buccal and sublingual administration in children.* Clin Pharmacokinet, 2006. **45**(7): p. 745-54.

284. Kokki, H., et al., *Pharmacokinetics of oxycodone after intravenous, buccal, intramuscular and gastric administration in children.* Clin Pharmacokinet, 2004. **43**(9): p. 613-22.

285. Zin, C.S., et al., *A randomized, controlled trial of oxycodone versus placebo in patients with postherpetic neuralgia and painful diabetic neuropathy treated with pregabalin.* J Pain. **11**(5): p. 462-71.

286. Zin, C.S., et al., *An update on the pharmacological management of post-herpetic neuralgia and painful diabetic neuropathy.* CNS Drugs, 2008. **22**(5): p. 417-42.

287. Czarnecki, M.L., et al., *Controlled-release oxycodone for the management of pediatric postoperative pain.* J Pain Symptom Manage, 2004. **27**(4): p. 379-86.

288. Villa, M.P., et al., *Nocturnal oximetry in infants with cystic fibrosis.* Arch Dis Child, 2001. **84**(1): p. 50-54.

289. Balfour-Lynn, I.M., *Domiciliary oxygen for children.* Pediatr Clin North Am, 2009. **56**(1): p. 275-96, xiii.
290. Cachia, E. and S.H. Ahmedzai, *Breathlessness in cancer patients.* Eur J Cancer, 2008. **44**(8): p. 1116-23.
291. Currow, D.C., et al., *Does palliative home oxygen improve dyspnoea? A consecutive cohort study.* Palliat Med, 2009. **23**(4): p. 309-16.
292. Saugstad, O.D., *Chronic lung disease: oxygen dogma revisited.* Acta Paediatr, 2001. **90**(2): p. 113-5.
293. Ross, J.R., et al., *A systematic review of the role of bisphosphonates in metastatic disease.* Health Technol Assess, 2004. **8**(4): p. 1-176.
294. Howe, W., E. Davis, and J. Valentine, *Pamidronate improves pain, wellbeing, fracture rate and bone density in 14 children and adolescents with chronic neurological conditions.* Dev Neurorehabil, 2010. **13**(1): p. 31-6.
295. Wagner, S., et al., *Tolerance and effectiveness on pain control of Pamidronate(R) intravenous infusions in children with neuromuscular disorders.* Ann Phys Rehabil Med, 2011. **54**(6): p. 348-58.
296. Ringe, J.D. and J.J. Body, *A review of bone pain relief with ibandronate and other bisphosphonates in disorders of increased bone turnover.* Clin Exp Rheumatol, 2007. **25**(5): p. 766-74.
297. Duncan, A.R., *The use of subcutaneous pamidronate.* J Pain Symptom Manage, 2003. **26**(1): p. 592-3.
298. Hain R and Jassal S, *Oxford handbook of paediatric palliative medicine.* 2010: Oxford University Press
299. Ward, L., et al., *Bisphosphonate therapy for children and adolescents with secondary osteoporosis.* Cochrane Database Syst Rev, 2007(4): p. CD005324.
300. Scottish Dental Clinical Effectiveness Programme. *Oral Health Management of Patients Prescribed Bisphosphonates: Dental Clinical Guidance.* 2011; April [Available from: www.sdcep.org.uk.
301. NICE Clinical Guideline. *Feverish illness in children. CG160.* . 2013; May [Available from: http://guidance.nice.org.uk/CG160.
302. Pillai Riddell, R.R., et al., *Non-pharmacological management of infant and young child procedural pain.* Cochrane Database Syst Rev, 2011(10): p. CD006275.

303. Uman, L.S., et al., *Psychological interventions for needle-related procedural pain and distress in children and adolescents.* Cochrane Database Syst Rev, 2006(4): p. CD005179.

304. Wong, I., C. St John-Green, and S.M. Walker, *Opioid-sparing effects of perioperative paracetamol and nonsteroidal anti-inflammatory drugs (NSAIDs) in children.* Paediatr Anaesth, 2013. **23**(6): p. 475-95.

305. Wong, T., et al., *Combined and alternating paracetamol and ibuprofen therapy for febrile children.* Cochrane Database Syst Rev, 2013. **10**: p. CD009572.

306. Rowland, A.G., et al., *Review of the efficacy of rectal paraldehyde in the management of acute and prolonged tonic-clonic convulsions.* Arch Dis Child, 2009. **94**(9): p. 720-3.

307. Ahmad, S., et al., *Efficacy and safety of intranasal lorazepam versus intramuscular paraldehyde for protracted convulsions in children: an open randomised trial.* Lancet, 2006. **367**(9522): p. 1591-7.

308. Armstrong, D.L. and M.R. Battin, *Pervasive seizures caused by hypoxic-ischemic encephalopathy: treatment with intravenous paraldehyde.* J Child Neurol, 2001. **16**(12): p. 915-7.

309. Giacoia, G.P., et al., *Pharmacokinetics of paraldehyde disposition in the neonate.* J Pediatr, 1984. **104**(2): p. 291-6.

310. Koren, G., et al., *Intravenous paraldehyde for seizure control in newborn infants.* Neurology, 1986. **36**(1): p. 108-11.

311. Appleton, R., S. Macleod, and T. Martland, *Drug management for acute tonic-clonic convulsions including convulsive status epilepticus in children.* Cochrane Database Syst Rev, 2008(3): p. CD001905.

312. Yoong, M., R.F. Chin, and R.C. Scott, *Management of convulsive status epilepticus in children.* Arch Dis Child Educ Pract Ed, 2009. **94**(1): p. 1-9.

313. www.palliativedrugs.com, *Phenobarbital* 2010.

314. Holmes, G.L. and J.J. Riviello, Jr., *Midazolam and pentobarbital for refractory status epilepticus.* Pediatr Neurol, 1999. **20**(4): p. 259-64.

315. Osorio, I., R.C. Reed, and J.N. Peltzer, *Refractory idiopathic absence status epilepticus: A probable paradoxical effect of phenytoin and carbamazepine.* Epilepsia, 2000. **41**(7): p. 887-94.

316. Bourgeois, B.F. and W.E. Dodson, *Phenytoin elimination in newborns.* Neurology, 1983. **33**(2): p. 173-8.

317. Tudur Smith, C., A.G. Marson, and P.R. Williamson, *Phenytoin versus valproate monotherapy for partial onset seizures and generalized onset tonic-clonic seizures.* Cochrane Database Syst Rev, 2001(4): p. CD001769.
318. Tudur Smith, C., et al., *Carbamazepine versus phenytoin monotherapy for epilepsy.* Cochrane Database Syst Rev, 2002(2): p. CD001911.
319. McCleane, G.J., *Intravenous infusion of phenytoin relieves neuropathic pain: a randomized, double-blinded, placebo-controlled, crossover study.* Anesth Analg, 1999. **89**(4): p. 985-8.
320. Mendoza, J., et al., *Systematic review: the adverse effects of sodium phosphate enema.* Aliment Pharmacol Ther, 2007. **26**(1): p. 9-20.
321. Miles C, F.D., Goodman ML, Wilkinson SSM., *Laxatives for the management of constipation in palliative care patients.* The Cochrane Collaboration.; The Cochrane Library. 2009: JohnWiley&Sons, Ltd.
322. El-Tawil, S., et al., *Quinine for muscle cramps.* Cochrane Database Syst Rev, 2010(12): p. CD005044.
323. MHRA. *Quinine: not to be used routinely for nocturnal leg cramps.* 2010; Available from: http://www.mhra.gov.uk/Safetyinformation/DrugSafetyUpdate/CON085085.
324. Bell, S.G., *Gastroesophageal reflux and histamine2 antagonists.* Neonatal Netw, 2003. **22**(2): p. 53-7.
325. Tighe, M.P., et al., *Current pharmacological management of gastro-esophageal reflux in children: an evidence-based systematic review.* Paediatr Drugs, 2009. **11**(3): p. 185-202.
326. Moayyedi, P., et al., *Pharmacological interventions for non-ulcer dyspepsia.* Cochrane Database Syst Rev, 2006(4): p. CD001960.
327. Wang, Y., et al., *Additional bedtime H2-receptor antagonist for the control of nocturnal gastric acid breakthrough.* Cochrane Database Syst Rev, 2009(4): p. CD004275.
328. Grassi, E., et al., *Risperidone in idiopathic and symptomatic dystonia: preliminary experience.* Neurol Sci, 2000. **21**(2): p. 121-3.
329. Kenrick S, f.S., *Treatment guidelines for symptom crises in Juvenile Battens Disease.* 2011.
330. BTS/SIGN. *British Guideline on the management of asthma. National clinical guideline.* 2012; May 2008 revised Jan 2012 [Available from: www.sign.ac.uk/pdf/sign101.pdf.

331. Chavasse, R., et al., *Short acting beta agonists for recurrent wheeze in children under 2 years of age.* Cochrane Database Syst Rev, 2002(3): p. CD002873.

332. Palazon Garcia, R., A. Benavente Valdepenas, and O. Arroyo Riano, *[Protocol for tizanidine use in infantile cerebral palsy].* An Pediatr (Barc), 2008. **68**(5): p. 511-5.

333. Henney, H.R., 3rd and M. Chez, *Pediatric safety of tizanidine: clinical adverse event database and retrospective chart assessment.* Paediatr Drugs, 2009. **11**(6): p. 397-406.

334. Vasquez-Briceno, A., et al., *[The usefulness of tizanidine. A one-year follow-up of the treatment of spasticity in infantile cerebral palsy].* Rev Neurol, 2006. **43**(3): p. 132-6.

335. Spiller, H.A., G.M. Bosse, and L.A. Adamson, *Retrospective review of Tizanidine (Zanaflex) overdose.* J Toxicol Clin Toxicol, 2004. **42**(5): p. 593-6.

336. Chauhan, S., et al., *Tranexamic acid in paediatric cardiac surgery.* Indian J Med Res, 2003. **118**: p. 86-9.

337. Frachon, X., et al., *Management options for dental extraction in hemophiliacs: a study of 55 extractions (2000-2002).* Oral Surg Oral Med Oral Pathol Oral Radiol Endod, 2005. **99**(3): p. 270-5.

338. Graff, G.R., *Treatment of recurrent severe hemoptysis in cystic fibrosis with tranexamic acid.* Respiration, 2001. **68**(1): p. 91-4.

339. Mehta, R. and A.D. Shapiro, *Plasminogen deficiency.* Haemophilia, 2008. **14**(6): p. 1261-8.

340. Morimoto, Y., et al., *Haemostatic management of intraoral bleeding in patients with von Willebrand disease.* Oral Dis, 2005. **11**(4): p. 243-8.

341. Fahn, S., *High dosage anticholinergic therapy in dystonia.* Neurology, 1983. **33**(10): p. 1255-61.

342. Ben-Pazi, H., *Trihexyphenidyl improves motor function in children with dystonic cerebral palsy: a retrospective analysis.* J Child Neurol, 2011. **26**(7): p. 810-6.

343. Rice, J. and M.C. Waugh, *Pilot study on trihexyphenidyl in the treatment of dystonia in children with cerebral palsy.* J Child Neurol, 2009. **24**(2): p. 176-82.

344. Hoon, A.H., Jr., et al., *Age-dependent effects of trihexyphenidyl in extrapyramidal cerebral palsy.* Pediatr Neurol, 2001. **25**(1): p. 55-8.

345. Tsao, C.Y., *Low-dose trihexyphenidyl in the treatment of dystonia.* Pediatr Neurol, 1988. **4**(6): p. 381.
346. Marsden, C.D., M.H. Marion, and N. Quinn, *The treatment of severe dystonia in children and adults.* J Neurol Neurosurg Psychiatry, 1984. **47**(11): p. 1166-73.
347. Sanger, T.D., et al., *Prospective open-label clinical trial of trihexyphenidyl in children with secondary dystonia due to cerebral palsy.* J Child Neurol, 2007. **22**(5): p. 530-7.
348. Brook L, V.J., Osborne C., *Paediatric palliative care drug boxes; facilitating safe & effective symptom management at home at end of life.* Archives of Disease in Childhood, 2007. **92 (Suppl I): A58**.
349. 3Dickman, A., J. Schneider, and J. Varga, *The Syringe Driver. Continuous Infusions in Palliative Care.* 2005: Oxford University Press.

Practical Handbook References

1. Field And Richard E. Behrman 'Communication, Goal Setting, And Care Planning' In When Children Die: Improving Palliative And End-Of-Life Care For Children And Their Families The National Academies Press 2001
2. Buckman R. How To Break Bad News: A Guide for Health Care Professionals. Baltimore, Md: Johns Hopkins University Press; 1992:15.
3. Bluebond-Langner, Myra *The Private Worlds Of Dying Children*, Princeton Paperbacks, Princeton University Press. 1980
4. Together for Short Lives care pathwayhttp://www.togetherforshortlives.org.uk/professionals/care_provision/care_pathways
5. See Together for Short Lives's Transition Care Pathway: http://www.act.org.uk/page.asp?section=115&sectionTitle=Together for Short Lives%27s+transition+care+pathway See Together for Short Lives's Transition Care Pathway: http://www.act.org.uk/page.asp?section=115&sectionTitle=Together for Short Lives%27s+transition+care+pathway
6. In . Tindyebwa, Kayita, Musoke et al (Eds) *African Network For Care Of Children Affected By HIV/AIDS (ANECCA) Handbook On Paediatric AIDS In Africa.* Revised Edition 2006. ....Www.Anecca.Org..
7. Together for Short Lives Care Pathway *http://www.act.org.uk/landing.asp?section=97&sectionTitle=Care+pathways+for+babies%2C+children+and+young+people)*
8. Larue F, Brasseur L, Musseault P, Et Al. 'Pain And Symptoms In HIV Disease: A National Survey In France.' Abstract: Third Congress Of The European Association For Palliative Care. *J Palliat Care* 10:95, 1994.
9. Foley F. 'AIDS Palliative Care'. Abstract: 10th International Congress On The Care Of The Terminally Ill. *J Palliat Care* 10:132, 1994.
10. Moss V. 'Palliative Care In Advanced HIV Disease: Presentation, Problems, And Palliation'. *AIDS* 4(S):S235-42, 1990.
11. Fontaine A, Larue F, Lassauniere JM. 'Physicians Recognition Of The Symptoms Experienced By HIV Children:How Reliable?' *J Pain Symptom Manage* 18:263-70, 1999.
12. Breitbart W, Mcdonald MV, Rosenfeld B, Monkman ND, Passik S. 'Fatigue In Ambulatory AIDS Children'. *J Pain Symptom Manage* 15:159-67, 1998.
13. Department For International Development (DFID) Www.Dfid.Gov.Uk/Mdg/Childmortalityfactsheet.
14. WHO World Health Report 2003 – Shaping The Future. WHO Geneva., Www.Who.Int/Whr/2003/Chapter1/En/Index2.Html Department For International Development (DFID)

15. Taken From "Conditions Suggestive Of Less Than Six Months Prognosis In Children With AIDS" *Recommendations For Managing Hiv Infection In Children*. Repubic of South Africa Department of Health, October 2001
16. Strafford M, Cahill C, Schwartz T, Et Al. 'Recognition And Treatment Of Pain In Pediatric Children With AIDS' *J Pain Symptom Manage* 6:146,1991.
17. Mathews W, Mccutcheon JA, Asch S, Et Al. 'National Estimates Of HIV-Related Symptom Prevalence From the HIV Cost And Services Utilization Study'. *Med Care* 38:762, 2000.
18. Reprinted With Permission From Hewitt D, Mcdonald M, Portenoy R, Et Al. 'Pain Syndromes And Etiologies In Ambulatory AIDS Children'. *Pain* 70:117-23, 1997
19. Amery J, Children's Palliative Care in Africa, OUP, Oxford, 2009
20. Eland, JM: 'Pediatrics'. In *Pain*. (1985) Springhouse, PA. Springhouse Corporation
21. Such as 'Wong Faces' (www.wongbakerfaces.org), the Faces Scale (www.usask.ca/childpain/fpsr), or the Visual Analogue Scale or (www.health.vic.gov.au/qualitycouncil/downloads/app1_pain_rating_scales).
22. For example the Eland body tool in 'Eland J, Anderson J. The experience of pain in children. In: Jacox A, editor. Pain: A Sourcebook for Nurses and Other Health Professionals. Boston: Little Brown; 1977. pp. 453–78.'
23. http://pain.about.com/od/testingdiagnosis/ig/pain-scales/Flacc-Scale.htm
24. http://www.bmj.com/content/315/7111/801.full) (See Part XXX Spiritual needs)
25. Texas Childrens Cancer Centre, Texas Childrens Hospital, Houston, Texas. Www.Childcancerpain.Org
26. Uman LS, Chambers CT, Mcgrath PJ, Kisely: Cochrane Review *'Psychological Interventions For Needle-Related Procedural Pain And Distress In Children And Adolescents'* Http://Www.Cochrane.Org/Reviews/En/Ab005179.Html.Http://Www.Cochrane.Org/Reviews/En/Ab005179.Html. October 18. 2006.
27. Von Baeyer CL, Marche TA, Rocha EM, Salmon K. 'Children's Memory For Pain: Overview And Implications For Practice'. *The Journal Of Pain* 2004;5(5):241-249
28. http://specialchildren.about.com/od/mentalhealthissues/a/breathing.htm
29. http://www.yourfamilyclinic.com/adhd/relax.htm
30. www.who.int/cancer/palliative/painladder/en/
31. Finkel JC, Finley A, Greco C, et al; Transdermal fentanyl in the management of children with chronic severe pain: results from an international study.; *Cancer*. 2005 Dec 15;104(12):2847-57.
32. From BNF for Children 2013-2014
33. Dunphy K et al. Rehydration in palliative and terminal care: if not, why not? Palliative Medicine 1995: 9: 221-8.

34. Burge FI. 'Dehydration And Provision Of Fluids In Palliative Care. What Is The Evidence?' *Canadian Family Physician.* 1996: 42:2383-8. (SA, 14 Refs)
35. Bruera, E., Belzile, M., Watanabe, S. & Fainsinger, R.L (1996). Volume of hydration in terminal cancer patients. *Supportive Care in Cancer,* 4 (2), 147-50.
36. Burge, F.I (1993). Dehydration symptoms of palliative care. *Journal of Pain and Symptom Management,* 8 (7), 454-464.
37. McCann, R.M., Hall, W.J. & Groth-Juncker, A. (1994). Comfort care for terminally ill patients. The appropriate use of nutrition and hydration. *JAMA,* 272 (16), 1263-6.
38. Bruera, E., Belzile, M., Watanabe, S. & Fainsinger, R.L (1996). Volume of hydration in terminal cancer patients. *Supportive Care in Cancer,* 4 (2), 147-50.
39. Chadfield-Mohr, S.M.,& Byatt, C.M. (1997). Dehydration in the terminally ill- iatrogenic insult or natural process? *Postgraduate Medical Journal,* 73 (862), 476-80
40. Fox, E.T. (1996). IV hydration in the terminally ill: ritual or therapy? *Br J Nurs,* 5 (1), 41-5.
41. Parkash, R. & Burge, F. (1997). The family's perspective on issues of hydration in terminal care. *Journal of Palliative Care,* 13 (4), 23-7.
42. Parkash R. Burge F. 'The Family's Perspective On Issues Of Hydration In Terminal Care'. *Journal Of Palliative Care. 13(4):23-7, 1997* (I)
43. Steiner, N. & Bruera, E (1998). Methods of hydration in palliative care patients. *Journal of Palliative Care,* 14 (2), 6-13.
44. Tindyebwa, Kayita, Musoke et al (Eds) *African Network For Care Of Children Affected By HIV/AIDS (ANECCA) Handbook On Paediatric AIDS In Africa.* Revised Edition 2006. ..Www.Anecca.Org.
45. Amery JM And Rose CJ 'Quantitative Evaluation Of Children's Palliative Care Service In Uganda'. *Cardiff Children's Palliative Care Conference* 2008 (In Press)
46. Management of Severe Malnutrition: A manual for physicians and other senior health workers. World Health Organization. Geneva 1999. ISBN 92 4 154511 9 . Online at www.who.int/nutrition/publications/malnutrition/en/index.html
47. Merriman A. *Pain And Symptom Control In The Cancer And/Or Aids Patient In Uganda And Other African Countries* Hospice Africa Uganda. P.O. Box 7757, Kampala ISBN 9970-830-01-0. Fourth Edition 2006
48. Pearl RH, Robie DK, Ein SH, et al. Complications of gastroesophageal antireflux surgery in neurologically impaired versus neurologically normal children. *J Pediatr Surg.* Nov 1990;25(11):1169-73
49. Renee Hsia, November 7, 2006, 'Pediatrics, Gastrointestinal Bleeding', *E-Medicine, Www.Emedicine.Com/Emerg/Topic381*
50. Www.Palliativedrugs.Com
51. http://www.mhra.gov.uk/Safetyinformation/DrugSafetyUpdate/CON300404
52. Clayden, GS: Management Of Chronic Constipation. Archives Of Disease In Childhood (1992) 67: 340-44

53. Meuser T. Pietruck C. Radbruch L. Stute P. Lehmann KA. Grond S. 'Symptoms During Cancer Pain Treatment Following WHO-Guidelines: A Longitudinal Follow-Up Study Of Symptom Prevalence, Severity And Etiology.' *Pain*. 2001: 93(3):247-57. (OS-593)
54. Sullivan, PB: 'Gastrointestinal Problems In The Neurologically Impaired Child'. *Bailliere's Clinical Gastroenterology'*. (1997) 11,3 529-46
55. Merriman A. *Pain And Symptom Control In The Cancer And/Or Aids Patient In Uganda And Other African Countries* Hospice Africa Uganda. P.O. Box 7757, Kampala ISBN 9970-830-01-0. Fourth Edition 2006
56. Tindyebwa, Kayita, Musoke et al (Eds) *African Network For Care Of Children Affected By HIV/AIDS (ANECCA) Handbook On Paediatric AIDS In Africa*. Revised Edition 2006. Www.Anecca.Org. Pp 95
57. Gwyther, Merriman, Mpanga Sebuyira, Schietinger. *A Clinical Guide To Supportive And Palliative Care For HIV/AIDS In Sub-Saharan Africa* Foundtaion For Hospices In Sub-Saharan Africa. 2006. Www.Fhssa.Org
58. WHO/UNICEF *Joint Statement On Clinical Management Of Acute Diarrhoea*, May 2004
59. Alexander CS. Palliative And End Of Life Care. In Anderson JR, Ed. *A Guide To The Clinical Care Of Women With HIV*. Rockville, MD: U.S. Dept. Of Health And Human Services Health Resources And Services Administration, 349-82, 2001. In Gwyther, Merriman, Mpanga Sebuyira, Schietinger. A Clinical Guide To Supportive And Palliative Care For HIV/AIDS Chapter 6: Pulmonary Symptoms
60. Gwyther L, Adams V, Wilson D, Mandwa D. 'Chapter 6: Pulmonary Symptoms' In Gwyther, Merriman, Mpanga Sebuyira, Schietinger. *A Clinical Guide To Supportive And Palliatuve Care For HIV/AIDS In Sub-Saharan Africa*. Foundtaion For Hospices In Sub-Saharan Africa. 2006. Www.Fhssa.Org
61. British Thoracic Society guidelines for assessment of persistent cough in children http://www.brit-thoracic.org.uk/Portals/0/Guidelines/Cough/Guidelines/cough_in_children.pdf
62. Reuben DB, Mor V, Hiris J 1988. 'Clinical Symptoms And Length Of Survival In Patients With Terminal Cancer'. *Arch Int Med* 148(7), 1586
63. Alexander CS. 'Palliative And End Of Life Care'. In Anderson JR, Ed. *A Guide To The Clinical Care Of Women With HIV*. Rockville, MD: U.S. Dept. Of Health And Human Services Health Resources And Services Administration, 349-82, 2001.
64. Gwyther, Merriman, Mpanga Sebuyira, Schietinger. *A Clinical Guide To Supportive And Palliative Care For HIV/AIDS In Sub-Saharan Africa*. Foundtaion For Hospices In Sub-Saharan Africa. 2006. Www.Fhssa.Org. Pp125
65. Hartshorne ST. 2000. 'Common Dermatological Problems Among HIV/AIDS Patients'. *CME: YOURSA Journal Of CPD*. 18:321–326.

66. Tuthill J, Garnier S. 2003. 'Prevention Of Skin Breakdown'. In Gwyther, Merriman, Mpanga Sebuyira, Schietinger. Eds. *A Clinical Guide To Supportive And Palliative Care For HIV/AIDS.* Foundtaion For Hospices In Sub-Saharan Africa. 2006. Www.Fhssa.Org
67. Merriman A. *Pain And Symptom Control In The Cancer And/Or Aids Patient In Uganda And Other African Countries* Hospice Africa Uganda. P.O. Box 7757, Kampala ISBN 9970-830-01-0. Fourth Edition 2006
68. Kandel, E.R., J.H. Schwartz, and T.M. Jessell. 1991. *Principles of Neural Science,* 3rd edition. New York: Elsevier.d
69. Hartley, Frank and Goldenson *Understanding Children's Play*, Routledge, 1999
70. 'National Institute for Play': www.nifplay.org
71. Child development Insititute: www.childdevelopmentinfo.com
72. Schore, A.N. The seventh annual John Bowlby memorial lecture. Minds in the making: attachment, the self-organizing brain, and developmentally-oriented psychoanalytic psychotherapy. *British Journal of Psychotherapy*,17, 299-328
73. Sheets-Johnstone, Maxine, (1999) The Primacy of Movement, Johns-Benjamin Vol. 14, *Advances in Consciousness Research*
74. Opie. I. (1993) *The People in the Playground*. New York, Oxford Univ. Press
75. Pelligrini, A.D. (1988) Rough-and-Tumble play from childhood through adolescence. In D. Fromberg & D. Bergen (Eds.) *Play from birth to twelve and beyond: Contexts, perspectives, and meanings.* (pp. 401-408). New York, Garland.
76. Paley, V.G. (1992) *You Can't Say You Can't Play.* Cambridge MA, Harvard University Press
77. Frank Wilson, (1999) *The Hand: How Its Use Shapes the Brain, Language, and Human Culture (*Vintage)
78. Singer, Dorothy G. with Singer, Jerome L. *The House of Make-Believe: Children's Play and the Developing Imagination.* 1990
79. Winnicott, D. W. *Playing and Reality.* London: Routledge, 1999.
80. Evans, M. (1993), 'Teenagers And Cancer', In *Paediatric Nursing*, 1993, Vol. 5. No 1. P 14-15
81. Weller B *Paediatric Nursing And Techniques*, London. Harper Row. 1985
82. Mackenzie H 'Teenagers In Hospital'. *Nursing Times*. 1988, 84: 32. 58-61
83. Evans, M. (1993), 'Teenagers And Cancer', In *Paediatric Nursing*, 1993, Vol. 5. No 1. P 14-15
84. Denholm C, 1987. 'The Adolescent Patient At Discharge And The Post-Hospitalisation Environment' *Maternal Child Nursing Journal* 1987. 16:95-101
85. Viner R And Keane M, 1998. Youth Matters: Evidence-Based Best Practice For The Care Of Young People In Hospital. Pubs: Caring For Children In The Health Services. London.
86. Stevens M. 'Care of the dying child and adolescent - family adjustment and support' in Doyle, Hanks Cherney and Calman,(Eds) *Oxford Textbook of Palliative Medicine.* Third Edition OUP 2005

[87] Stevens M. 'Care of the dying child and adolescent - family adjustment and support' in Doyle, Hanks Cherney and Calman,(Eds) *Oxford Textbook of Palliative Medicine*. Third Edition OUP 2005

[88] Module 7: Working with Adolescents in AIDS Relicf and Catholic Relief Services: *Psychosocial Care & Counselling for HIV Infected Children and Adolescents: A Training Curriculum.* Catholic Relief Services 228 W. Lexington St. Baltimore, Maryland 21201-3413 | 888-277-7575 | info@crs.org. January 2008 (in press)

[89] Blomquist KB, Brown G, Peersen A And Presler EP . 'Transitioning To Independence: Challenges For Young People With Disabilities And Their Caregivers' *Orthopaedic Nursing* 17(3) 27-35 1998

[90] Stevens SE Et Al 'Adolescents With Physical Disabilities: Some Psychosocial Aspects Of Health'. *Journal Of Adolescent Health* 1997: 19 (2) 157-64

[91] UNICEF, UNAIDS, and WHO. 2002. *Young People and HIV/AIDS: Opportunities in Crisis.* New York: UNICEF

[92] Njovana E, Watts C. 'Gender Violence In Zimbabwe: A Need For Collaborative Action'. *Reprod Health Matters,* 1996; (7):46-52.

[93] Bohmer L. 'Adolescent Reproductive Health In Ethiopia: An Investigation Of Needs, Current Policies And Programs. Los Angeles, Ca: Pacific Institute Of Women's Health, 1995

[94] WHO World Health Report 2003 – Shaping The Future. WHO Geneva., Www.Who.Int/Whr/2003/Chapter1/En/Index2.Html Department For International Development (DFID)

[95] Wjau W, Radeny S. *Sexuality Among Adolescents In Kenya.* Nairobi: Kenya Association For The Promotion Of Adolescent Health, 1995.

[96] Relief and Catholic Relief Services: *Psychosocial Care & Counselling for HIV Infected Children and Adolescents: A Training Curriculum.* Catholic Relief Services 228 W. Lexington St. Baltimore, Maryland 21201-3413 | 888-277-7575 | info@crs.org. January 2008 (in press)

[97] Carr-Gregg M, Sawyer S, Clarke C And Bowes G 'Caring For The Terminally Ill Adolescent'. *MJA* Vol. 166, 3rd March 1997, Pp 255-258

[98] Corr C and Balk D *Handbook of Adolescent Death and Bereavement.* Springer Publishing Company (April 1996)

[99] Papadatou, D. 1989. Caring For Dying Adolescents. Nursing Times

[100] Papadatou, D. 1989. Caring For Dying Adolescents. Nursing Times May 3rd, Vol. 85: 18, 28-31

[101] See http://www.ncbi.nlm.nih.gov/pubmed/11648226

[102] http://www.bma.org.uk/ethics/consent_and_capacity/mencaptoolkit.jsp.

[103] See http://www.patient.co.uk/doctor/Consent-to-Treatment-in-Children.htm

[104] Children's Palliative Care Guidelines. Royal College of Paediatrics and Child Health. London

[105] Stoter D (1995) *Spiritual Aspects of Health Care.* London

[106] Khaneja and B. Milrod, 'Educational needs among pediatricians regarding caring for terminally ill children'. *Arch Pediatr Adolesc* Med 152 1998

[107] GMC Treatment and care towards the end of life: good practice in decision making. **http://www.gmc-uk.org/guidance/ethical_guidance/6858.asp** Together for Short Lives, 2011. *A Parent's Guide: Making critical care choices for your child*. Bristol: Together for Short Lives.

[108] See http://www.schloss-hartheim.at/index.asp?seite=560

[109] Gold Standards Framework http://www.goldstandardsframework.nhs.uk/OneStopCMS/Core/CrawlerResourceServer.aspx?resource=2E9DA6B2-5D8D-4D3B-8315-C0FA90181045&mode=link&guid=4913518d1abe4b2d976d500829f54c5b

[110] Download an example of an Advance Care Plan developed by South Central SHA (LINK to http://www.southcentral.nhs.uk/wp-content/uploads/2010/09/Child-and-Young-Persons-ACP-Form.doc)

[111] Royal College of Paediatrics and Child Health (RCPCH), 2004. *Withholding or Withdrawing Life Saving Medical Treatment in Children: A framework for practice*, 2nd Edition. London: Royal College of Paediatrics and Child Health.

[112] http://www.rcn.org.uk/__data/assets/pdf_file/0009/270873/4.7.2_Palliative_care_drug_boxes.pdf)

[113] Worden, J W. Grief Counselling and Grief Therapy: A Handbook for Mental Health Practitioners Springer, .New York 2009

[114] Foxhall, Zimmerman, Standley And Ben. 'A Comparison Of Frequency And Sources Of Nursing Job Stress Perceived By Intensive Care, Hospice And Medical-Surgical Nurses'. *Jounal Advanced Nursing* 1990 15:577-84

[115] Bene And Foxhall. ' Death Anxiety And Job Stress In Hospice And Medical-Surgical Nurses'. *Hosp Journal* 1991 7:25-31

[116] Woolley, Stein, Forrest, Baum. 'Staff Stress And Job Satisfaction At A Childrens Hospice'. *Arch Disease In Childhood*. 1989. 64:114-118

[117] Amery, J, Lapwood, S 'A Study Into The Educational Needs Of Children's Hospice Doctors' *Palliative Medicine* 18(8) 727-733, 2004 (Dec)

[118] Papadatou D, Healthcare Providers' Responses To The Death Of A Child In Oxford Textbook Of Palliative Care For Children Oxford University Press. Oxford 2006

[119] Amery And Lapwood (See Iv)

[120] Dunwoodie; Auret, 'Psychological morbidity and burnout in palliative care doctors in Western Australia', *Internal Medicine Journal*, Volume 37, Number 10, October 2007, pp. 693-698(6)

Main Index

Adherence 2, 31, 145–146, 149
Adjuvants 40, 49–51, 53, 56, 58–59, 77
Adolescence 141, 144–146, 405
Advanced Care Plan 168, 173
Aggression, Aggressive 94, 135, 165
Agitation 34, 42, 53, 62–63, 78, 105, 124, 129, 171
Aids 30–34, 37, 42–43, 59, 67, 69, 87, 90, 108, 112–113, 115–117, 145–146, 149, 401–406
Anaemia 38, 40, 129
Analgesia 49, 51–52, 57–58, 106
Anorexia 33, 38, 64
Anticholinergic 58, 67, 70, 72–73, 83–84, 106
Anticipatory 50, 183
Anticonvulsant 41–42
Antidepressants 36, 41–42, 84
Antiemetic 38, 53, 78–79
Antihistamine 36, 41, 84, 92, 111–112
Anti-Inflammatory 57, 72–73, 113
Antiretroviral 32, 35, 87
Antispasmodic 57, 85, 89
Anxiety, Anxiolytic, Anxious 12, 33–34, 44, 49, 58, 77–78, 81–82, 92, 104–105, 118–121, 128, 130–131, 133, 143, 163, 171, 183, 191–193, 195–196, 407
Aphthous Ulcer 39
Appetite 65–66
Art 8, 13, 30–32, 35–36, 38, 41–42, 88, 92, 98, 100, 114–116, 121, 138, 144, 146, 148–149, 159
Asphyxia 103
Aspiration 64, 69–72, 92, 96, 101, 145
Attachment 405
Autonomy 146

Behaviour 6, 8, 13, 46, 51, 120–121, 144, 180, 191, 193
Belief 20, 121, 153–154, 157, 161, 170
Bereavement, Bereaved 32, 63, 179–183, 185, 406
Bleeding 57, 74–75, 93–95, 103, 124, 128–130, 171, 403
Breathlessness, Breathless 40, 65, 91–92, 94, 96–98, 127, 130–131
Bronchiectasis 40
Bronchodilator 91, 98–99
Bronchospasm 91, 96–98
Burnout 187–190, 192, 195, 407
Cachexia 65
Cancer xvii, xix, 4, 8, 55–56, 59, 66, 69, 108, 145, 149, 151, 402–405
Candida 69
Capacity 156–157, 174, 406
Cardiomyopathy 30, 33
Cardiopulmonary Resuscitation 170
Catharsis, Cathartic 17
Catheter 59, 172
Chaplain 25
Chemotherapy 42, 50, 56, 59, 66, 69, 79, 88, 98
Closure 179
Cognition, Cognitive, Cognitive-Behavioural 51, 77, 121, 135, 138, 144
Colic 89
Collusion 4, 11–13, 25
Communication xviii, 1–5, 8–10, 12–14, 16–17, 22, 44, 119, 142–144, 147–148, 154, 159, 174, 188, 401
Community xiii, xvi, 15, 23, 26, 105, 158, 164, 179
Compassionate 62, 155, 185
Competence 150–151

409

Complementary 157
Confidentiality 24
Confusion 19, 25, 52
Consciousness 130, 191, 405
Conversations 4, 19, 137, 194
Coordination 22, 26
Cor Pulmonale 30, 96, 98
Cough 34, 40, 70, 80–81, 90–91, 93–95, 404
Counselling 28, 40, 42, 121, 179, 196, 406–407
Cremation 178–179
Cryptococcal 34, 42
Culture 2–3, 135, 162, 177, 179, 183, 405
Cytochrome P450, CYP 35–36
Cytotoxic 67–68, 80, 82
Death 1, 3–5, 9, 18–19, 22, 30, 32–33, 36, 50, 61, 73, 87, 93–94, 120, 123, 125–129, 131, 135, 144, 146, 148–149, 155, 162, 164–167, 169–171, 174–183, 185, 189, 406–407
Decisions 4, 10–11, 16, 23, 25, 32, 61, 94, 129, 142, 144, 148, 150, 152–156, 164–166, 168, 170–171, 174, 407
Dehydration 40, 61–62, 67, 83, 89, 403
Delirium 42, 129
Dependence 142, 145, 147
Depression 4, 34, 37, 42, 147, 191–193, 195–196
Dermatitis 41, 111, 113–114
Diarrhoea 33, 37, 39, 65, 87–89, 404
Diet 39–40, 73, 81, 83–84, 88
Disability 65, 133, 196, 406
Disorientation 53
Distancing 4
Do Not Resuscitate, DNACPR, DNR 168–171, 174
Drama 148–149
Drooling 70, 104, 106
Drowsiness 37, 53

Dying xi, 3, 7, 18, 31–32, 65, 120, 122, 124, 144, 147–148, 158, 160–162, 177–178, 180–181, 185, 189–190, 401, 405–406
Dysmotility 69, 72, 78, 81, 87–88, 104
Dysphagia 39, 69–70, 74, 92
Dyspnoea 34, 44, 96
Dystonia 57, 59, 72, 83, 103–107
Education xvi–xvii, xix, 28–29, 31, 143, 145–146, 156, 186
Effusion 40, 97, 99
Embalm 175
Emergencies, Palliative 123
Emotion 4–5, 13, 15, 17–18, 20, 135, 143, 147, 180–183, 189
Empathy 154, 188
Encephalopathy 30, 34, 41, 84, 103
End-Of-Life Care 163–168, 174, 401, 404
End Of Life Care Plan 167–168
Enema 83, 85
Environment 5–6, 11, 18, 28, 49, 105–106, 119, 134–135, 142, 144, 169, 405
Epidural 56
Epilepsy 104
Epistaxis 130
Ethics 62–63, 126, 151, 154, 156, 406
Ethnic, Ethnicity 24, 55, 161
Euphemisms 19
Euthanasia 126, 128
Exhaustion 33, 154, 180
Faith 158
Family, Families xiii, xv–xviii, 1, 3–5, 8–15, 17–18, 22–29, 31–33, 37, 45–47, 49, 62–63, 65, 71, 75, 85, 88, 93–94, 96, 100–102, 104–105, 111–112, 119–121, 123–129, 133, 135, 141–142, 145, 147–149, 157–159, 163–168, 170–172, 175–183, 185, 193, 401, 403, 405–406

410

Fatigue 34, 38, 94, 103, 401
Fear 4–5, 10, 12, 18, 25, 46, 49, 51, 77, 83, 120, 126, 133, 142–143, 146–147, 155, 160, 165, 188
Feed, Feeding 34, 38–39, 59, 61–67, 70–73, 85, 88–89, 92, 104–105
Feeling 4, 6, 11, 14–15, 65, 120, 124–126, 141, 144, 146, 148–149, 158, 160, 180, 182–183, 188–189, 191
Fever 33, 42, 88, 90–91, 93–94, 96, 100, 104
Fits xxi, 13, 101, 126, 148, 170, 176
Flatulence 81
Fluids 38, 46, 61–63, 70, 75, 85, 87, 89, 95, 99, 131, 403
Folliculitis 41, 108, 112–113
Formulary ix, xi, xviii, xxi, 52
Friends 5, 146, 159, 167, 169, 178, 181–182, 185, 190–192
Funeral 147, 169, 177, 183–184
Fungal, Fungating 41, 108, 112–114, 116
Gag 70, 77, 80–81, 92
Gastritis 35, 57, 74–75, 81
Gastrostomy 66, 71
Gingivitis 34, 39
God 157, 160
Grandparents 27, 169, 180
Grief, Grieve, Grieving 146, 175, 178–183, 185–186, 188, 407
Guilt 25, 32–33, 126
Gut 77
Haematemesis 73, 93
Haemoptysis 93–95
Haemorrhage 129–130
Haemostatic 95
Hallucination 129
Headache 33–34, 81
Heaven 19, 160
Hepatitis 32, 35
Hernia 72
Herpes 39–41, 67–68, 108, 115–116

HIV, AIDS xiv, 30–35, 37–38, 42–43, 59, 67, 69, 84, 87–88, 90, 99, 108, 112–117, 145–146, 149, 401–406
Holistic xiii, 24, 133, 147
Honesty 10, 20, 153
Hospices ix, xiii–xix, 16, 26, 101, 122, 125, 132, 151, 403–405, 407
Hydrotherapy 105
Hypercalcaemia 84
Hyperglycemia 35
Hyperhidrosis 108, 117
Hypersalivation 72
Hypersensitivity 41, 88
Hypnosis 51
Hypodermoclysis 63
Hypoventilation 96
Hypoxia 94, 100, 103
Identity 22, 142, 157
Imagination 19, 51, 125, 136, 405
Immunocompromise 68, 115–116
Impaction 53, 88–89
Incontinence 61, 88
Independence 25, 142–143, 145, 406
Infection 31, 33–35, 38–41, 58–59, 64–65, 67, 69–70, 72, 80–82, 87–91, 93–94, 96, 98, 103, 105, 108, 110–115, 117, 129, 136, 402
Infestations 41, 108, 112
Inflammation 39, 55, 57, 69, 90, 114
Informed Consent 152
Insomnia xiii
Intolerance 87–88
Intractable Pain 158–159
Irritability 188
Isolation 77, 141, 147, 158
Itch 111
Jaundice 111
Language 2–3, 10, 19, 24, 47, 135, 138, 405
Larygomalacia 90
Laxative 40, 53, 83, 85–86, 88
Leukoencephalopathy 33

Life-Sustaining Treatment 155, 167–168, 170
Life-Threatening xv, 12, 17, 29, 33, 74, 76, 93, 133, 150, 165, 173
Listening 1–2, 5, 15, 120–121, 159, 180, 195
Loneliness 159, 188
Loss 5, 33, 38, 61–62, 72, 87, 91, 94, 130, 142–143, 146, 159, 178–181, 183, 185–186, 188
Lying 73, 97, 101, 121, 177, 184
Lymphoma 33, 42, 108, 117
Malabsorption 39, 87
Malignancy 33, 38–39, 41, 91, 93–94
Malnutrition 42, 64–65, 69, 88, 403
Massage 51, 59–60
Memory, Memorial xiv, 14, 58, 77, 148, 169, 181, 183, 185, 402, 405
Meningitis 34, 42
Mindfulness 6
Mobility 42, 84, 103, 105
Molluscum 108, 115
Mood 46, 141, 191
Mortality 5, 177
Mortuary 175, 177
Motility 70–71, 81, 89
Multi-Disciplinary 22–23, 26, 153
Musculoskeletal 59
Music 75, 120–121, 136–139, 144, 148–149
Myth 5, 56
Narrative 135
Nasogastric Tube 63
Nausea 33–34, 38, 44, 53, 58, 64–66, 77–78, 80–81, 84, 104, 128, 171
Neuralgia 41
Neurodisability 84
Neurology, Neurological 56, 59, 64, 69–70, 79, 81, 83–85, 88, 100
Neuromuscular 64, 69, 96
Neuropathy, Neuropathic 30, 34, 49, 56–59

Neurotransmitter 57, 103
NMDA 57
Nociceptors 49
Non-Pharmacological Treatment 104
Non-Verbal communication 1, 5, 14, 159
Normalise, Normalisation 119–120
Nutrition 38, 42, 59, 88, 403
Obsessive-Compulsive 196
Obstructive 111
Oedema 61, 63, 97–98, 131
Oesophagitis 74–75, 81
Oncology xiii, xvi, 15, 21, 56, 71
Opioid xvii, 40, 49, 51–54, 57, 66, 80, 83–84, 92, 95–96, 117, 171
Opportunistic Infection 31, 34–35, 38
Oxygen 90, 92, 94, 97–101, 130
Pain xiii, xix, 4–5, 14, 28, 30–31, 33–34, 38–41, 43–60, 64–65, 67, 70–71, 74–75, 77, 81, 83–85, 87–88, 94, 96–97, 103–107, 115, 122, 124, 126, 128–129, 143, 157–159, 164, 169, 171, 181, 184, 188, 190, 401–405
Pain Sieve 45
Palpitations 37
Pancreatitis 35
Panic 14, 75, 100, 124
Papules 112, 115
Parent 2–4, 6–8, 10–12, 16, 31, 45–48, 61, 65, 135, 142–143, 145–147, 152–155, 158–159, 165, 168, 170–171, 178–181, 185, 407
Pastoral Care 157–158, 162
Patch, Patches xviii, 54–55, 113, 125, 131
Peace 121, 147, 167
Permission 117, 147–148, 402
Physiotherapy 40, 60, 91, 99, 105
Platelet 75, 95, 125, 129–130
Play 2–3, 28, 30, 42, 48, 87, 121, 133–140, 148–149, 154, 179, 183, 192, 405

Pleurodesis 99
Pneumonitis 30, 90–91, 99
Pneumothorax 97–98
Poetry 138, 148–149
Positioning 59, 65, 73, 91
Post-Mortem 174–175
Pre-Bereavement 182
Pressure Sore 42
Proctitis 89
Proctoclysis 63
Prognosis, Prognostication 10, 25, 27, 32–33, 62, 65, 94, 98, 141, 147–148, 165, 402
Prophylaxis 53, 115–116
Psoriasis 113–114
Psychology, Psychological ix, 9, 24–25, 27, 31, 49, 58, 61, 77, 135, 141, 143, 147, 154, 166, 188–191, 402, 407
Psychosocial 62, 406
Pyrexia 55
Radiotherapy 56, 59, 76, 82, 88, 92, 95, 98, 115–116
Rapport 5, 7, 144, 147
Rash 37, 108, 113
Reflux 64, 69–70, 72–74, 80–81, 90, 92, 105
Regurgitation 80–81
Rehydration 39, 63, 89, 402
Relationship 26, 143–144, 147–148, 158, 181, 189–190
Relax, Relaxation 5, 40, 49, 59, 86, 97, 121, 402
Religion, Religious 19–20, 28, 154, 157, 161, 169–170, 180
Resilience 182, 188–189
Respect 3, 6–7, 24, 142, 144, 150, 179
Respiratory 37, 40, 90, 93, 96, 98, 171
Restlessness 47, 53, 129
Resuscitation 164, 169–171, 173
Retention 52–53, 90, 105–106, 129
Rhythm 137

Rights 4–5, 7, 11, 16, 18, 29, 64, 78, 139, 150–154, 158, 160, 191–192, 195
Rituals 175–176, 179, 403
Sadness 17, 180, 182–183, 187
Scabies 108, 111–112
Scapegoating 188–189
Scenario Planning 125, 127, 155, 163
School xiv–xv, xviii–xix, 26, 29, 31, 46, 119–120, 143, 145–146, 159, 183
Secretion 61, 70, 85, 87, 91, 104, 131, 171
Sedation, Sedative 37, 56–57, 85, 95, 106, 128–130
Seizure 72, 83, 100–102, 104–105, 124–125, 128, 172
Self-Esteem 145–146
Sepsis 42
Sex 145, 177
Sibling 27–28, 145, 147–148, 158–159, 169, 180–181, 185
Sleep 19, 46, 48, 56, 72–73, 138, 183
Soul xiii, 158, 162
Spasm 60, 69, 86, 103, 105–107
Spastic, Spasticity 57, 59, 103–107
S.P.I.K.E.S 15
Spinothalamic 49
Spiritual, Spirituality 20, 24–25, 27–28, 49, 58, 77, 141, 143, 147, 157–162, 164, 166, 169, 402, 407
Status 154
Steroid 38–42, 57, 59, 68–70, 72–73, 75, 81, 89, 91–92, 99, 111–114
Stigma 49
Stomatitis 67–68
Story, Stories 135, 137–138, 158–160, 178, 183
Stress 10, 13, 33, 59, 133, 135, 170, 188–189, 195, 407
Stridor 90, 96–97
Supervision 191
Support ix, xiv–xvi, 1, 6, 8–10, 15, 18, 26, 28, 32, 38, 44, 62, 97, 119,

129–130, 142–143, 145–146, 149, 155, 164, 169, 177, 179–180, 182, 184–186, 191, 193, 196, 405–406
Swallowing 64–65, 67, 130
Sweat 33, 42, 94
Symptom Management Plan 127, 163, 168, 171, 173
Syringe-Driver 54
Tachycardia 37, 62, 94
Tachypnoea 62, 94, 96
Teenager 120, 149, 405
Thalamus 49, 103
Therapy, Therapies, Therapist 25, 28, 30, 32, 35, 38, 41–42, 50, 68, 87, 114, 119, 121, 133, 144, 148, 403, 407
Tinea 114
Tolerance 102
Toxoplasmosis 33, 42
Tracheostomy 98
Tradition 158, 177, 179
Transdermal 51, 54, 402
Transfusion 38, 40, 95, 129, 178
Transplant 175–176
Trauma 41, 127, 135, 181, 188–189

Truth 12
Twitch 53
Ulcer 39, 68, 116–117
Undertaker 175
Urinary 52–53, 105, 129
Urticaria 112
Varices 74, 95, 129, 171
Vesicle 41, 68, 115
Vestibular 77, 81
Vomiting 34, 37–38, 53, 64, 67, 74, 77–82, 84–85, 93, 128, 171
Vulnerable 61, 110, 133, 145
Weakness 32, 96, 187, 190, 192
Wheeze 94
Who ix, xiii–xiv, xix, 2–4, 7–8, 12, 19, 23, 26–27, 35, 38, 43, 46–47, 49–51, 54, 56, 62, 64–65, 69, 72, 86, 106, 115, 121–122, 125, 133, 135–136, 142, 145, 147, 149, 151–153, 157–160, 167, 171, 174–176, 180–181, 184–185, 190–192, 401–404, 406
Withdraw 65, 148
Worry 19, 46, 160

Formulary Index

Acupuncture 346, 348, 366, 369
Adherence 387, 391
Adjuvants 232, 246, 253, 262, 264, 267, 280, 292, 312, 318, 346, 380, 384
Adolescence 380, 391
Adrenaline 202
Aggression, Aggressive 223, 328
Agitation 216–217, 233, 236–237, 260, 268, 274, 280, 292–293, 303–304, 306, 318–319, 371
Aids 379, 383
Anaemia 244, 270, 353
Analgesia 203, 214, 234, 244, 248, 250, 260, 271, 275, 283, 285, 288, 310, 316, 378, 384–385, 390, 392–393
Anorexia 208, 303–304, 394–395
Anticholinergic 289, 340, 399
Anticonvulsant 214, 220–222, 321, 346
Antidepressants 238, 252–253, 382
Antiemetic 230, 232–233, 262, 274–276, 289, 297, 307, 378–379, 391, 394–395
Antihistamine 238, 324
Anti-Inflammatory 232–233, 246, 265, 270, 299–300, 384, 397
Antimuscarinic 230, 324
Antipsychotic 218, 225, 238, 259, 304, 328, 394
Antispasmodic 339
Anxiety, Anxiolytic, Anxious 220, 222–223, 236–237, 252, 268, 275, 280, 282, 292–293, 335, 348, 353, 362, 371, 382
Apnoea 228
Appetite 206, 304

Aprepitant 207
Arachis Oil Enema 208
Arrhythmia 243, 247, 381
Arthritis 215, 239, 246, 265, 299, 377
Ascites 302, 353
Aspirin 209
Asthenia 208, 247, 276
Ataxia 254
Attachment 205
Behaviour 252
Bisacodyl 212
Bleeding 202, 245–246, 270, 302, 334, 338, 359–361, 365–366, 399
Bradycardia 286
Breathlessness, Breathless 266, 292, 310–311, 329, 353, 355, 396
Bronchodilator 330
Bronchospasm 247, 266, 329, 353
Cancer 207, 232, 261, 295, 376–378, 380–384, 387, 389–390, 392–396
Cardiovascular 216, 218, 239, 246, 265, 270, 300, 304, 328, 379
Chemotherapy 207, 215, 289, 297, 307–308, 375–376, 378
Chloral hydrate 216–217
Closure 239, 265
Codeine Phosphate 227–229, 241, 279, 337, 379
Cognition, Cognitive, Cognitive-Behavioural 238, 389
Colic 272, 288, 339, 346
Community 229, 388
Complementary 346, 348, 365, 368
Confusion 258–260, 286
Consciousness 244, 282
Cough 228, 252, 294, 395
Cryptococcal 251

Cyclo-Oxygenase Inhibitors 216, 239, 246, 265
Cytochrome P450, CYP 205, 238, 245, 247, 276, 287
Cytotoxic 232, 297
Death 243, 284–286, 381, 383
Decisions 311
Dehydration 373
Delirium 263, 303–304, 328, 371, 383, 394
Dependence 286
Depression 205, 214, 218–219, 228, 252, 281, 284, 286, 293, 297–298, 324, 338, 377, 382–383
Dermatitis 311
Diarrhoea 208, 228, 232, 244, 247, 252–253, 255, 273, 278–279, 288, 302, 387
Disability 388–389
Dizziness 208, 224, 230, 232, 247, 254, 270, 273, 300, 340
Drooling 222, 339–340
Drowsiness 206, 230, 232, 238, 262, 281, 286, 324, 336
Dying 221, 238
Dyspepsia 208, 247, 254, 270, 298, 305, 327
Dyspnoea 232, 234–235, 280, 292, 294–295, 396
Dystonia 327, 339–340, 362, 364, 376, 391, 398–400
Dysuria 268
Education 350
Effusion 353
Encephalopathy 271–272, 397
End-Of-Life Care 275, 319, 344
Enema 208, 241–242, 318, 323, 332–333, 373, 398
Enteral Feeding 206, 210–211, 215, 222, 225, 231, 234, 240, 244, 252, 257, 260–262, 266, 276, 278, 281, 290, 296, 306, 310, 316, 322, 328, 331, 333, 335–336, 338, 375
Entonox (nitrous oxide) 244
Environment 221, 371
Epidural 225
Epilepsy 211, 218–221, 293, 304, 318–320, 324, 328, 350, 398
Etoricoxib 246
Extrapyramidal 243, 304, 328, 399
Family, Families 350, 359
Fatigue 230, 247, 253–254
Fear 371
Feed, Feeding 206, 210–211, 215, 222, 225, 231, 234, 240, 244, 252, 257, 260–262, 266, 276, 278, 281–282, 290, 296, 300–301, 305–306, 310, 316, 323, 325, 327–328, 331, 333, 335–336, 338, 375
Feeling 330
Fever 218, 246, 254–255, 300, 316–317, 350, 373
Fits 350
Flatulence 247, 272–273, 279, 288
Fluids 247, 255, 282, 300, 313, 366, 373
Formulary 199–202, 375
Fungal, Fungating 250–251, 290–291, 301
Gag 282
Gastritis 247, 266, 326
Gastrostomy 238, 247, 274, 279, 292–293, 296, 305, 335, 340
Glycopyrronium bromide 256
Gums 291, 359, 365–366
Gut 272, 333, 348
Haematuria 268, 339
Haemorrhage 245–246, 341
Haemostatic 366, 399
Hallucination 258, 268, 286, 328
Headache 208, 224, 230, 232, 247, 252–253, 270, 273, 300, 308, 375, 377

Heartburn 247, 254
Herbal Treatment 346, 348, 366, 369
Hernia 385
Hiccup 208, 217–218, 258–259, 289, 377–378, 391
HIV, AIDS 287, 382–383
Hospices 244, 377, 381, 384, 390
Hypercalcaemia 312–313
Hyperkalaemia 329
Hypersalivation 256–257, 263
Hypersecretion 222
Hypersensitivity 246, 270, 299–300
Hyperthyroidism 211
Hypnosis 346, 348, 365, 368
Hypnotic 238, 389
Hypocalcaemia 313–314
Hypoglycaemia 350
Hypotension 205, 224, 276, 286, 336, 392
Hypothermia 205, 286
Hypothyroidism 259
Hypoxia 371
Ileus 241, 391
Immunocompromise 250–251, 291
Impaction 208, 281–282, 385
Incontinence 226–227, 271, 279, 388
Infection 218, 250–251, 254, 291, 301, 350, 359, 382
Inflammation 238, 246, 265, 300, 346
Insomnia 216–217, 233, 253, 377, 388–389
Intolerance 272
Irritability 223
Jaundice 218
Jejunostomy 238
Juvenile 215, 239, 265, 299, 328, 377, 398
Laxative 212, 215, 226–227, 282, 287–288, 298, 331–333, 373–374, 398
Leukaemia 379
Life-Threatening 229
Lomotil® (co-phenotrope) 278

Loss 244, 255, 288, 325, 360, 368
Malabsorption 314
Malignancy 359, 392
Mania 252, 303–304, 327
Massage 340, 346, 348, 362, 365, 368, 373
Methylnaltrexone 287
Mobility 368
Motility 242, 244, 380
Muscular Dystrophy 383
Musculoskeletal 239, 246, 264, 299
Music 371
Myasthenia 218
Nabilone 297
Nasogastric Tube 322
Nausea 207, 211, 217–218, 224, 230, 242, 247, 252–253, 258–259, 270, 272–273, 276, 279, 286, 288–289, 295, 297, 302–304, 307–308, 324–325, 327, 340, 348–349, 360, 371, 375–376, 379–380, 385, 387, 394
Neonate 203, 205, 216, 220–221, 228, 234, 236–237, 243, 245, 248, 251, 254, 256, 264–265, 271, 289, 291–295, 298, 301, 305, 311, 314–316, 318–321, 326, 329, 341, 351, 384–386, 392, 397
Neuralgia 387, 395
Neuroblastoma 384
Neurodegenerative 362, 380
Neurodevelopmental 389
Neurology, Neurological 214, 222, 289, 312–313, 340, 350, 368, 373, 376, 392, 396–399
Neuromuscular 205, 250, 312, 396
Neuropathy, Neuropathic 206, 209, 214–215, 220–221, 252–254, 267–268, 276, 283, 320, 383–384, 387, 392, 395, 398
NMDA 267–268
Obsessive-Compulsive 382

Obstructive 228
Oedema 247, 270, 286
Oesophagitis 247, 272, 282, 306, 334, 387
Off-Label Drugs 377
Oncology 375, 381–382, 390–393
Opioid 202–204, 211, 213–214, 222–224, 227–229, 234, 236, 241, 248–249, 260–261, 267–268, 272, 278, 281, 283–288, 294–295, 297–299, 307, 309–310, 337, 346, 377, 379–382, 384, 389–393, 395, 397
Oromucosal 248, 250, 293
Oropharyngeal 251, 291, 382, 391
Overdose 214, 278, 285, 287, 297–298, 316, 325, 399
Oxygen 244, 267, 310–312, 350–351, 353, 365, 396
Pain 201–202, 204, 206, 209, 212–215, 220–223, 227–228, 232, 234–236, 238, 240–241, 246–248, 250, 252–254, 260–262, 264–265, 267–270, 273–276, 283–285, 288, 292, 294–296, 299, 309, 312–314, 316, 318, 320, 337, 346–348, 353–354, 359, 362, 365–369, 371, 375, 377, 380–385, 387, 389–398
Palpitations 247, 286
Panic 220–221, 236–237, 292, 359
Paradoxical Pain 219
Parent 215, 353, 371, 388
Patch, Patches 213–214, 223, 225, 248–250, 263–264, 276–277, 356, 387
Pharmacokinetic 205, 225, 246, 250, 283, 286, 305, 377, 385–387, 392–393, 395, 397
Physiotherapy 362
Platelet 296, 338, 360, 378
Play 353

Pneumothorax 244
Positioning 340, 353, 356, 362, 371, 391
Pray, Praying, Prayer 346, 348, 359, 365, 368
Preterm 265, 291, 311, 315–316, 392
Prophylaxis 301, 305, 334
Pruritus 270, 295, 307, 395
Psychology, Psychological 346, 397
Psychosis, Psychotic 258, 303, 327–328
Psychosocial 368
Pyrexia 209, 214, 264, 314, 316
Radiotherapy 307, 368
Rash 252, 270, 273
Reflexology 346, 348, 365, 368
Reflux 242–243, 247, 254, 272, 282, 305–306, 326–327, 371, 380–381, 386–387, 391, 395, 398
Relationship 204, 321
Relax, Relaxation 346, 348, 359, 362, 365, 368
Respiratory 205, 214, 218, 221–222, 228, 237, 281, 284, 286, 293, 297–298, 311, 324, 338, 356, 358, 377
Restlessness 258–259, 371
Retention 208, 211, 247, 286, 295, 300, 340, 346, 371, 390
Rhythm 282, 388
School 350, 380
Secretion 230, 256–257, 262–263, 273, 356, 358
Sedation, Sedative 219, 221–224, 238, 263, 274–275, 280, 284, 292–293, 304, 318, 324, 359–360, 377, 379, 381
Seizure 220–222, 254, 292–293, 304, 318–321, 350–352, 371, 376, 380, 391–392, 397–398
Sepsis 371
Sleep 206, 224, 228, 268, 282–283, 324–325, 335, 379, 388–389

Spasm 231, 236–237, 262, 280, 286, 290, 327, 336, 346, 362, 371, 380
Spastic, Spasticity 209, 223–224, 231, 336, 376, 379–380, 399
Spiritual, Spirituality 359
Status 220–221, 237, 280, 292, 313, 318–321, 379–380, 389, 397
Steroid 305, 326, 366
Stomatitis 270
Stress 334
Suicide 252
Supervision 206, 248, 255, 286, 297, 328
Support 313, 368, 378, 380, 383, 390, 393, 395
Swallowing 274, 291, 301
Symptom Management Plan 345
Syringe-Driver 214, 222, 344, 389, 400
Tachycardia 286, 336, 340
Teenager 376
Therapy, Therapies, Therapist 210–211, 219, 232–233, 242, 253, 261–262, 282, 285, 297, 311–312, 317, 319, 322, 353, 381, 384–385, 388, 390–394, 396–397, 399
Tics 224
Tinnitus 270
Tolerance 221, 267, 331, 363, 384, 396
Tonic-Clonic 220, 320, 397–398

Transdermal 213, 223, 225, 248–249, 263, 356, 377, 381–382
Transfusion 360
Trauma 387
Truth 388
Ulcer 211, 247, 270, 272, 305, 326–327, 334, 359, 365–366, 393, 398
Urinary 211, 226–227, 268, 286, 295, 340, 368, 371, 376, 390
Urticaria 273
Varices 302, 334
Ventilation 202–203, 298, 392
Vertigo 286, 300, 324
Vestibular
Vitamin K (Phytomenadione) 341
Vomiting 207, 211, 217–218, 224, 230, 242, 247, 252, 255, 258–259, 273, 276, 286, 289, 295, 297, 302–304, 307–308, 324–325, 327, 348–349, 360, 373, 375–376, 379–380, 385, 387, 391, 394
Weakness 232, 329, 336, 368
Wheeze 399
Who 201, 218, 227–229, 236, 248, 251, 254, 292, 294, 308, 317, 322, 337, 340, 346, 350, 368, 375, 392–393
Withdraw 328, 340

Drug Index

Alfentanil 57, 202–205, 231, 375
Amitriptyline 36, 41, 56, 60, 84, 118, 206, 287, 346, 375
Amphotericin 68
Arachis 83, 208
Atropine 89, 278
Baclofen 57, 60, 106, 209–210, 363, 376, 379
Barbiturates 238
Benzene 112
Benzodiazepine 37, 40, 42, 57, 60, 73, 92, 95, 97–98, 106, 118, 121, 129, 219, 318, 362, 378, 392
Benztropine 106
Bethanechol 211, 376
Bisphosphonate 50, 313–314, 396
Botulinum 363, 381
Buprenorphine, Butrans 54, 57, 212–213, 346, 377
Buscopan 131
Calcipotriol 114
Carbamazepine 35, 41, 57, 207, 214, 287, 305, 322, 377, 397–398
Cefalosporins 99
Celecoxib 215–216, 377
Chloral 37, 216–217
Chloramphenicol 99
Chloroquine 36
Chlorpheniramine 111–112
Chlorpromazine 217, 378, 383
Clobazam 219
Clonazepam 220, 222
Clonidine 222–225, 378–379
Clotrimazole 39, 113–114
Co-Danthramer 226
Co-Danthrusate 226–227
Codeine 50, 55, 91, 227–229, 241, 279, 337, 379

Corticosteroid 57, 114, 207, 233, 270
Cortisol 382
Cotrimoxazole 66, 89
Cyclizine 78–79, 82, 172, 230, 271, 344, 348
Danthron 227
Dantrolene 106, 231, 363, 379
Dexamethasone 57, 66, 71, 78, 82, 98, 172, 207, 231–234, 346, 369–370, 379
Diamorphine 54–55, 60, 125, 128, 172, 203, 205, 229, 231, 234–236, 240, 249, 271, 343–344, 346, 359–360, 380
Diazepam 41, 57, 85, 97–98, 101–102, 106, 128, 222, 236, 238, 280, 292–293, 350–351, 360, 362–363, 371, 380, 388, 391
Diclofenac 238–239, 246, 266, 300, 346
Digoxin 98, 273, 325
Dihydrocodine 50, 55, 240
Diltiazem 36
Diphenhydramine 111
Diphenoxylate 89, 278
Docusate 83, 226–227, 241
Domperidone 71, 73, 78, 82, 242–243, 348, 380–381
Dopamine 78, 103, 106
Erythromycin 36, 71, 73, 113, 243, 245, 381
Etamsylate 245
Fentanyl 54–55, 57, 202, 204, 236, 248–250, 346, 375, 381–382, 385, 390, 402
Flucloxacillin 99, 111–112
Fluconazole 36, 39, 68, 70, 114, 250–251, 287, 366

Fluoxetine 36, 252–253, 382
Fluvoxamine 305
Furosemide 98, 131
Gabapentin 41, 56, 253, 383
Gaviscon 73, 81, 255
Glycerine 83, 85
Glycopyrrolate 118, 383
Griseofulvin 114
Haloperidol 36–37, 42, 78, 82, 129, 172, 258–259, 271, 344, 348, 371, 383
Heliox 92, 98
Hydromorphone 55, 260–261, 343, 384
Hyoscine Butylbromide 58, 89, 262
Hyoscine Hydrobromide 78, 172, 257, 262–263, 344, 356, 383
Ibuprofen 228, 239, 247, 264–266, 300, 317, 346, 397
Indometacin 239, 266, 300
Ipratropium 91, 266, 330
Isoniazid 36
Isphagula 83
Itraconazole 36, 39, 114, 207, 273
Ketamine 267–268, 347, 384
Ketoconazole 36, 113–114, 243
Ketorolac 239, 247, 266, 269–270, 300, 384–385
Lactulose 53, 83, 85, 271–272, 373, 385
Lansoprazole 73, 272–274, 386–387
Levomepromazine 78, 82, 121, 125, 129, 172, 271, 274, 276, 344, 348, 371, 387
Lidocaine 71, 276–277, 387
Lignocaine 91–92, 276, 387
Loperamide 39, 89, 279, 388
Lorazepam 57–58, 82, 101, 119, 125, 280–281, 354, 383, 388, 397
Macrogols 272, 281–282
Mannitol 42, 83
Megestrol 66, 395
Melatonin 282, 382, 388–389

Methadone 37, 54, 57, 283–287, 343, 389–390
Methotrexate 114, 300
Methotrimeprazine 304, 387
Metoclopramide 71, 73, 78–79, 103, 230, 244, 289–290, 348, 378, 380, 391
Metronidazole 36, 40, 42, 66–67, 89, 113, 116, 290, 391
Miconazole 291, 366, 391
Midazolam 41, 85, 95, 97, 101–102, 125, 128, 172, 271, 292–293, 319, 327, 344, 351, 354, 359–360, 362, 371, 378, 380–381, 391–392, 397
Morphine vi, 39–41, 50, 52, 54–55, 57, 60, 91, 97–98, 116, 125, 128, 202–204, 213–214, 227, 229, 234, 236, 240, 248–250, 260–261, 271, 285, 294–296, 298, 309–310, 337–338, 343–344, 346, 353, 359, 376, 380–382, 384, 389–390, 392–393
Naloxone 53, 214, 286–287, 297–299, 393
Naltrexone 286
Naproxen 239, 266, 299–301, 377
Nitrous Oxide 244, 347, 381
Nystatin 40, 67, 70, 301
Octreotide 85, 95, 302
Olanzapine 303–305, 394–395
Omeprazole 36, 38, 71, 73, 274, 305
Ondansatron 348
Oxycodone 52, 54–55, 57, 231, 271, 309–310, 343, 346, 381, 395
Pamidronate 312, 396
Paracetamol 50, 67, 228, 314, 316–317, 346, 384, 397
Paraffin 117, 290, 385
Paraldehyde 41, 101–102, 318, 351, 397

Phenobarbitone, Phenobarbital 35, 41, 101–102, 207, 221, 287, 318–320, 392, 397
Phenytoin 35, 41, 101–102, 221, 287, 318, 320–323, 397–398
Pholcodine 228–229
Phosphate 83, 85, 227, 229, 233–234, 323, 398
Phytomenadione 341
Piroxicam 239, 247, 266, 270, 300
Poloxamer 83, 226
Prednisolone 35, 68, 233
Pregabalin 41, 57, 395
Prolactin 304
Promazine 42, 78
Promethazine 112, 324–325
Propranolol 118
Quinine 36, 325–326, 398
Ranitidine 57, 70–71, 73, 326–327
Rifampicin 207, 287
Risperidone 327–328, 398

Salbutamol 91, 329–330
Senna 83, 85, 331, 373
Septrin 36–37, 99
Sodium Citrate 332
Sodium Picosulphate 83
Sorbitol 83, 210, 244, 332
Sucralfate 274, 334–335
Temazepam 335
Terbinafine 114
Terfenidine 36
Tetrabenzene 106
Tizanidine 106, 336, 363, 399
Tramadol 54–55, 337
Tranexamic Acid 75–76, 95, 130, 338, 360, 366, 399
Trihexyphenidyl 339–340, 399–400
Trimeprazine 388
Trimethoprim/Sulfamethoxazole 36–37
Verapamil 36
Warfarin 75, 207, 247, 300, 325